I0082104

A Nurse's 'Comforts'

The Diary and Letters of Louisa Stobo AANS, WW1

Elizabeth Butel

ETT IMPRINT

Exile Bay

First published by ETT Imprint, Exile Bay 2025

ETT IMPRINT
PO Box R1906
Royal Exchange NSW 1225 Australia

 ISBN   978-1-923205-75-8

Design by Tom Thompson

Cover: Louisa Stobo, second from right, with friends and fellow nurses at the Sphinx in 1915. Collection of Dinah Toohey.

For families who have lost loved ones in war,
especially the family of Robert Scobie.

Abbreviations

AAH   Australian Auxiliary Hospital
AAMC Australian Army Medical Corps
AANS Australian Army Nursing Service
ACF   Australian Comforts Fund
ARC   Australian Red Cross
AGH   Australian General Hospital
AIF   Australian Imperial Force
ANTA Australasian Trained Nurses Association
ASH   Australian Stationary Hospital
ACCS Australian Casualty Clearing Station
AWM Australian War Memorial
BEF   British Expeditionary Force
HMAT His Majesty's Australian Transport
N.D.   No date

The full words for abbreviations used by Louisa in both letters and diary, which may cause confusion, are given in brackets after the abbreviation.

CONTENTS

Louisa Scobie on her graduation from Sydney Hospital. [1]

1 Collection of Dinah Toohey

Preface

On 25 November 1914, Louisa Stobo (née Scobie) set sail from Sydney with other members of the Australian Army Nursing Service (AANS), only months after war was declared. She had been a member of the AANS since May 1907, but this was her first active service. She was among the earliest of the many trained nurses who rushed to enlist[2] and that November, she boarded the *Kyarra,* the first Australian hospital ship to leave. The Australasian Nurses' Journal referred to her as "foremost amongst nurses from Sydney" in their description of military nurses bound for Egypt.[3]

Louisa was the seventh child and sixth daughter of a notable Maitland family of fifteen children. Her parents, Robert[4] and Mary Scobie had been scions of the local district at their Oakhampton farm, 'The Gardens,' where fruit trees grew, and table grapes were cultivated.

When she left Sydney, Louisa was two years married. In September 1910, she had resigned from Sydney Hospital,[5] where she trained and nursed for eleven years, to marry Edye Stobo, a land agent of 14 Castlereagh Street, Sydney. Stobo was a widower, 20 years her senior, with a grown family of five sons. Marriage was meant to preclude continuing work as a nurse, but nonetheless, her embarkation papers (25.11.1914)[6] cite her as Matron of Crown Street Women's Hospital and married to Edye Stobo. Records from the hospital state that she was given leave of absence to serve.[7]

Louisa was born on 16th January 1876 and was a mature woman of nearly 40 when she set out. Theoretically, nurses who enlisted were required to be unmarried and aged 25 to 40. But Louisa, among many others, just fitted the age restriction and since her attestation papers (generally referred to as her 'papers'[8] in the remainder of this text) clearly show her as married, the requirement that women must be single, was also waived.

2.   Harris, K., 2014. 'New horizons: Australian nurses at work in World War I', Endeavour, Vol.38, No.2, p.1
3.   The Australasian Nurses' Journal, Vol. XII, No. 12, December 1914, pp.398-403
4.   Robert Scobie Senior had represented the Hunter Electorate in the New South Wales Legislative Assembly, from 1889-1893
5.   Sydney Hospital was Sydney's first and oldest hospital, where nurses were trained in the Nightingale method, stressing adequate treatment, layout and sanitation, ventilation and nourishing food
6.   National Archives of Australia, NAA: B2455, Stobo Louisa https://recordsearch.naa.gov.au/ SearchNRetrieve/Interface/ViewImage.aspx?B=8092348
7.   The Medical Journal of Australia, 29 May 1915
8.   See details above in note 6.

In these 'papers', she is described as a woman of 5 feet 8 inches (c.172 cm), weighing 185 pounds (c.84 kg), with a dark complexion, brown hair and brown eyes.[9] On the voyage out she kept a diary[10] that gives some insight into her state of mind as she travelled towards these life-changing events. The skills and experience she took with her would be tested in the many roles she would perform in Egypt, France, Malta and England, so a brief description of them may shed light on her preparedness for the years ahead.

Louisa Scobie trained at Sydney Hospital in the early years of the 20th Century, winning a prize in 1901 for efficiency in the theory and practice of nursing for her first year of study in 1900.[11] Over the training period, she attended lectures in anatomy, physiology, medical and surgical nursing and invalid cooking, supplemented by practical demonstrations and instruction by matrons and senior nurses. These years saw Sydney Hospital develop new methods of treatment, with massage added in 1902, ophthalmic nursing in 1903, and for a period, lectures on mental health nursing that same year.[12] Louisa would likely have lived on site, in the Nightingale Wing, which had been upgraded in 1901 for the use of the Matron, nursing staff and personal attendants. Similarly, a new observation ward had been added to the Casualty department in 1903. Five years later, hygiene and pediatric nursing were added to the training curriculum.[13]

If, as it is believed, Louisa stayed at Sydney Hospital for eleven years,[14] this was in excess of the six years nursing staff were generally retained at the hospital.[15] According to a later account of her service there,[16] she held the positions of senior nurse, sister and for a period, acting matron. The long-term Matron of Sydney Hospital in that era, Rose Creal, testified to hers and the Medical Superintendent's "entire satisfaction" with how Louisa had performed her duties.[17] This was high praise, for Creal was regarded as a strict

9. The attestation papers incorrectly describe her as 8 feet 8 inches in height
10. The diary is from the collection of Sandra Roberts
11. 'A Perfectly Managed Institution', Daily Telegraph, Wednesday 27 February 1901, page 4
12. Watson, J. F. (1911) The History of the Sydney Hospital from 1811 to 1911, Government Printer, Sydney, 1911, p. 178
13. ibid
14. 'Matron of the Hospital, Sister Stobo Appointed. A Varied Experience,' Maitland Daily Mercury, Thursday 11 August 1921, page 2
15. 'Nursing as a Profession,' Sunday Times, Sydney, NSW, Sunday 21 February, p. 7
16. 'Matron of the Hospital, Sister Stobo Appointed', The Maitland Mercury, Thursday, 11 August 1921, page 2
17. ibid

disciplinarian who, nonetheless, won respect and admiration for the care she showed to others.[18]

Sydney's *Sunday Times* published an extensive interview with Matron Creal in February 1904, which gives more insight into the sort of training Louisa received.[19] Initial applicants for training had to be of good height and physique, good character and general fitness. They were accepted as probationers for a three-month course, including household economics, diet sheets, personal and general hygiene, disinfection and sterilisation. If satisfactory, they began formal studies, progressing from first year anatomy, physiology and general nursing to second year studies in medical, surgical and ophthalmic nursing. In the third year, they studied invalid cookery, massage and electricity,[20] and general nursing, and in the fourth year, dispensing, housekeeping, and optional midwifery nursing. Examinations took place at the end of each year, with additional examinations in practical work in the fourth year of study. Fourth year nurses earned £39 per year, head nurses, £52, sisters, £72 and senior sisters, £84.

Louisa's absence from nursing was brief. At her marriage to Edye Stobo, late in January 1911, her occupation is listed as 'home duties'.[21] But by mid-January 1913, she had been appointed Matron of The Women's Hospital, Crown Street, Surry Hills.[22] Newspaper accounts of the appointment mention that she received all her training at Sydney Hospital and Melbourne Women's Hospital, the latter referring to a course in hospital administration that Louisa may have undertaken, in preparation for her new role at Crown Street.

Historian Judith Godden's excellent history of Crown Street Women's Hospital[23] records the transformations that took place there, shortly after Louisa started her appointment. In April 1913 the Albion Street extension to the hospital was opened, lit by electric lights.

18.   MacDonnell, F. (1981) Rose Ann Creal (1865-1921), Australian Dictionary of Biography, Volume 8, 1981
19.   'Nursing as a Profession,' Sunday Times, Sydney, Sunday 21 February 1904, p. 7
20.   Ruffin, P. T., (2011), A History of Massage in Nurse Training School Curricula (1860-1945), Journal of Holistic Nursing, American Holistic Nurses Association, Vol. 29, No. 1 March 2011, pp. 61-67. Ruffin notes that Florence Nightingale had promoted the use of massage in the 1860s. She also notes that electrical devices were used to treat disease by electrotheraputists
21.   New South Wales Marriage Certificate, Registration Number: 1911/002013
22.   'Matron appointed', Truth, Sydney, Sunday 19 January 1913, page 8
23.   Godden, J. (2017) Crown Street Women's Hospital: A History 1893-1983, Allen & Unwin

The new building had lifts, reducing the need for nurses to physically carry seriously ill patients up several flights of steps. Despite these welcome changes, the hospital, located in one of Sydney's then poorest districts, was chronically short of funds and needed many more beds and a new laundry to improve standards of accommodation and hygiene.[24]

The year Louisa started as Matron, the hospital treated over 3,000 patients, most of whom were considered impoverished. We may assume that her duties were similar to those of the previous founding Matron, Hannah McLeod, who was in charge of managing nursing, assisting doctors to teach trainees and organizing district midwifery.[25] As Godden describes, it was a rigorous institution where nurses were dismissed for insubordination or breaking rules. Highly regarded and described as 'the mother of the institution',[26] McLeod had died prematurely in October 1912, leading to the search for a new Matron and Louisa's subsequent appointment.

Godden suggests that Louisa may have added to her training with a midwifery course funded by a legacy from her father,[27] but that otherwise, she was not a likely choice, as there is no record of that training. When she was appointed in January 1913, reports described the hospital as having 45 beds, with a majority constantly occupied[28] but 80 additional beds envisaged.[29] In the meantime, hundreds of cases were reputedly turned away for lack of space.[30] Louisa is quoted as distressed at the cases she had to refuse.[31]

In mid-1913, she was taking reporters on a tour of the hospital and revealing her need to practise stringent economy in order to provide patients with nourishing food. One newspaper described how Matron Stobo, along with other nurses, visited municipal vegetable markets to buy food wholesale.[32]

24. ibid, p. 25
25. ibid, p. 35
26. 'Personal', Daily Telegraph Sydney, Monday 28 October 1912, page 8
27. Godden, Crown Street Women's Hospital: A History 1893-1983, p. 44
28. 'Of Interest to Women', Sydney Mail, Wednesday 22 January 1913, page 23
29. 'Personal', The Sydney Morning Herald, Friday 17 January 1913, page 8
30. 'Queen Anne and Mary Ann, an Appeal to Save Life', Daily Telegraph, Sydney, Friday 20 June 1913, page 9
31. ibid
32. 'Alms and the Women, Mothers in Hospital', Crown Street, Sun, Wednesday 2 July 1913, p 10

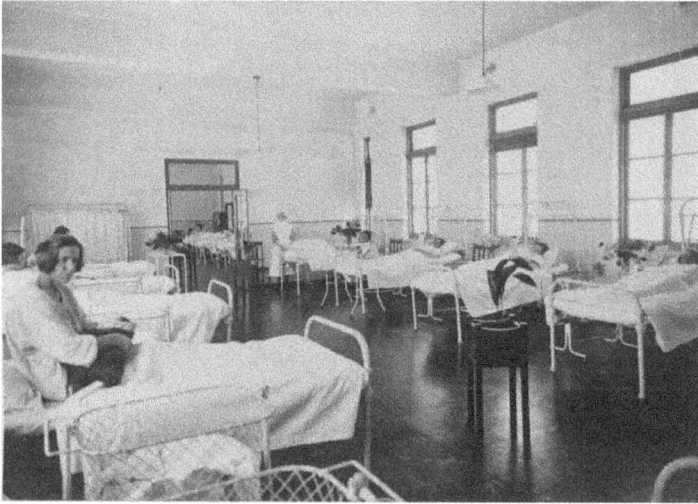

Crown Street Women's Hospital, Maternity Ward,
City of Sydney Archives, Sydney Reference Collection, A-00028718

Publicity was necessary because the hospital's overdraft was at tipping point and an appeal, and a benefit had been organized to raise funds. According to one report,[33] the hospital was over £7000 in debt, with the upkeep of a bed costed at £76 a year. Patients, for their part, were only able to contribute an average of 4 shillings for their own care. As well, a certain stigma was attached to the hospital, coming from the ill-founded belief that most of the patients were unmarried.[34] In any case, admission to the hospital was based on need, and women's marital status had no impact on the decision to admit them. Rather poverty was "the entrance ticket."[35]

By October that year, the money had been raised and the crisis was past. A lessening of pressure may be inferred when, along with staff from other hospitals, Louisa and selected nurses were invited to join a lunch at Admiralty House, to honour visiting dignitaries.[36]

---

33.   'For Women in the Throng', Daily Telegraph, Thursday 19 June 1913, page 8
34.   'The Women's Hospital', The Sydney Morning Herald, Wednesday, 18 June 1913, page 7
35.   'Queen Anne and Mary Ann, An Appeal to Save Life', Daily Telegraph, Sydney, Friday 20 June 1913, page 9
36.   'Luncheon at Admiralty House', Daily Telegraph, Monday 6 October 1913

One of Louisa's last engagements with Crown Street was a celebration where she entertained an array of invited luminaries to tea. The gathering was to celebrate the 21st anniversary of the founding of the hospital,[37] and the opening of a new obstetric ward, named in honour of the late Hannah McLeod.

Judith Godden notes that Louisa was one of the first staff members from Crown Street to volunteer[38] and that her replacement, Wilhelmina Sölling, who had also trained at Sydney Hospital, met with hostility when almost the whole of the midwifery staff resigned at her appointment. Godden suggests that the staff wanted an in-house appointment and protested at bringing in an outsider for a second time. She also records that, unlike the doctors who enlisted, neither Louisa, nor Sölling, who enlisted in 1915, had their positions at the hospital kept for them.

During the pre-war years, Louisa had also become a member of the Australasian Trained Nurses Association, (ATNA), which had been established in New South Wales in 1899, with other state branches set up over the next 8 years.[39] Its aim was to establish a system of registration for trained nurses, develop training standards, and, in general, improve the status of nurses.

Initiated in the larger training hospitals, it may be that Louisa joined on completion of her own training. Her commitment to the professional status of nurses is also shown by her membership of the Australian Army Nursing Service, which was set up in 1902.

Louisa joined the AANS in 1907, with her name appearing as a scrutineer in the election of their council[40] on at least two occasions.[41] According to the Women's Register,[42] the AANS was staffed by volunteer

37. 'An Anniversary,' Daily Telegraph (Sydney NSW:1833-1930) Saturday 17 October 1914
38. Godden, Crown Street Women's Hospital: A History 1893-1983, p. 48
39. Francis, R. & Carter, C. (2009), 'Australasian Trained Nurses' Association', The Australian Women's Register, https://www.womenaustralia.info/entries/australasian-trained-nurses-association/
40. 'Australasian Trained Nurses' Association, Election of Council,' The Sydney Morning Herald (NSW: 1842-1954), Friday, 24 July 1908, page 3
41. 'Australasian Trained Nurses' Association, Election of Council,' The Sydney Morning Herald (NSW: 1842-1954), Saturday, 24 July 1909, page 6
42. 'Australian Army Nursing Service' (AANS). (1902-1948). (2008). In Trove. Retrieved September 26, 2024, from https://nla.gov.au/nla.party-597790

civilian nurses and functioned as a reserve who would be available in times of national crisis. Members were required to be fully trained nurses with at least three years of medical and surgical nursing. When war broke out in 1914, medical staff who were willing to serve were recruited from the AANS and from the civilian nursing population. This long-term commitment to the AANS may have been the reason Louisa volunteered for overseas service at the first opportunity, despite her role as Matron at Crown Street and her marriage to Edye Stobo, only a few years before.

Table 1: Australian Army Nursing Service (AANS) 1914 - 15 outdoor dress.[44]
Nurses were given an allowance of £19.10.00 to equip themselves

1 grey skirt and blouson (jacket)
1 rainproof overcoat
1 grey cape
2 red shoulder capes
1 bonnet
2 embroidered AANS sleeve badges
2 brassards (armbands)
6 pairs starched cuffs
3 grey zephyr dresses (ward dress)
8 white linen aprons
3 white belts
1 pair black boots
Trimmings

1 grey cloak
1 pair grey gloves
2 sets Australian Military forces buttons
1 silver Rising Sun badge
6 starched collars
4 linen squares (veils x 1-yard^2)
4 grey aprons
3 grey belts
1 hold-all
3 Red Cross armbands

43. Harris, K. (2009). Red Reflections on the Sea: Australian Army Nurses Serving at Sea in World War 1. The Journal of Australian Naval History, 6(2), 51–73.https://search-informit org.ezproxy.lib.uts.edu.au/doi/10.3316/informit.261756961849214
44. Wadman, A. (May. 2014) 'Australian War Memorial', https://www.awm.gov.au/articles/blog/ australian-army-nursing-service-1914-15-outdoor-dress

Louisa's Diary

Louisa sailed from Sydney in late November 1914 on HMAT *Kyarra,* a cargo and passenger liner which had been converted into a hospital ship. Its destination was England. By the time the ship left Fremantle, there were several hospitals on board:

No. 1 Australian General Hospital (No. 1 AGH)

No. 2 Australian General Hospital (No. 2 AGH)

No. 1 Australian Stationary Hospital (No. 1 ASH)

No. 2 Australian Stationary Hospital (No. 2 ASH)

No. 1 Australian Casualty Clearing Station (No. 1 ACCS)

The Commonwealth Defence Department had a policy of recruiting hospital personnel for the Australian Imperial Force (AIF) from each state. No. 1 AGH was formed in Queensland, while No. 2 AGH was formed in New South Wales. The smaller No. 1 ASH was largely raised in South Australia, No. 2 ASH in Western Australia and No. 1 ACCS in Tasmania.[45]

The ship set sail in Brisbane, then docked in Sydney, Melbourne and Fremantle, taking on personnel at each stop. Personnel from South Australia and Tasmania boarded in Melbourne.[46]

These medical units were envisaged as part of a multi-stage process of treating the wounded.[47] Time was critical, particularly for soldiers with abdominal injuries, when intervention could reduce rates of infection, amputation and death.

An article in the Sydney Morning Herald (SMH)[48] gave extensive details of the number and kind of hospitals on board, stating that in total, they would provide 1640 beds for sick and wounded soldiers. Louisa was linked to the Sydney contingent and No. 2 AGH, under the leadership of Lieutenant-Colonel Thomas Morgan Martin.[49]

45. Barrett, J. W. & Deane, P. E. (1918), The Australian Army Medical Corps in Egypt, H. K. Lewis & Co., London

46. 'No. 1 Australian General Hospital', https://throughtheselines.com.au/research/1-AGH

47. '1918, Australia's medical personnel', Australian War Memorial, (2024) https://www.awm.gov.au/visit/exhibitions/1918/medical

48. 'Hospital Ship Kyarra, Australian Units,' Sydney Morning Herald (NSW : 1842 - 1954), Thursday 26 November 1914, page 8

49. Thomas Morgan Martin, born in 1854, was commanding officer of No. 2 Australian Army General Hospital

As well as medical personnel, there were carpenters, dispensers, a skiagrapher (radiographer), electro-therapy attendants, operating room orderlies, and cooks who specialised in invalid cookery. The medical personnel numbered c.83 medical officers, 180 nurses, and c.500 rank and file.[50] The *Kyarra* also carried ambulances, X-Ray equipment and motor vehicles.

The principal Matron on board was Jane Bell, who became a vigorous advocate for nurses in the Australian Army Medical Corps (AAMC), particularly regarding their authority and working conditions.[51] As historian Kirsty Harris points out, being a military nurse "significantly differed to the skills required to nurse in Australia"[52] and this, and other challenges, led to conflicts with some in the army who resented the presence of women so close to the battlefronts.

HMAT *Kyarra,* 1916 after it was converted to a troopship. AWM, C1218526.

50. Barrett and Deane, The Australian Army Medical Corps in Egypt, p. 13
51. Gardiner, L. (1979), 'Jane Bell', Australian Dictionary of Biography, Vol. 7 1979
52. Harris, K., 2014, 'New horizons: Australian nurses at work in World War I', Endeavour Volume 38 No. 2

Sister Anne Kidd-Hart, of the AANS, who was also on board the *Kyarra*, left her impressions of the voyage in an interview in March 1919. She recalled that the ship was "frightfully crowded and the trip over was not very comfortable."[53] This view is supported by the history of the AAMC, published in 1918, which stated that the ship was not fit for purpose and was small, overcrowded and unsanitary. The *Kyarra* was slow, the engines broke down and the food caused ptomaine poisoning at one stage.[54]

The ship was docked for several days in Melbourne when it was found to be carrying ordinary cargo, leaving insufficient space for the necessary medical equipment. The cargo was removed and replaced but there was still a lack of space for the Red Cross goods attached to No. 1 AGH.[55] Under the protection of the Red Cross, the *Kyarra* was permitted to carry only medical units and hospital equipment.[56]

The journey to Alexandria took over seven weeks and during that time, those on board required mandatory vaccinations against smallpox and typhoid. Australia's official war historian, Charles Bean, believed that these vaccinations, an initiative of Colonel Neville Howse,[57] assistant director of medical services, 1st Australian Division, contributed to the Australian forces' ability to hold the trench lines.

The first AIF contingent, part of an Australasian convoy of Australian and New Zealand ships, left Albany on 1st November 1914. The Australian contingent included medical personnel as well as a principal matron, three other matrons, and c.20 sisters - senior members of Nursing Staffs of Nos. 1 and 2 AGH.[58] On seven of the troop transports, nurses trained orderlies, ambulance personnel and members of medical detachments, as well as assisting medical officers with vaccinations and innoculations.[59]

53. Sister Anne Kidd-Hart was born 1875 in Victoria. She was one of a large number of nurses interviewed about their experiences while awaiting transport back to Australia in 1919. AWM 41: 1072. Interviews containing accounts of Nursing experiences in the AANS (Australian Army Nursing Service), p. 45
54. Barrett and Deane, The Australian Army Medical Corps in Egypt, p. 15
55. ibid, p. 14
56. 'Hospital Steamer *Kyarra*, Contraband on Board,' Brisbane Courier, Tuesday 5 January 1915, p 7
57. Braga, S. (2000), *Anzac Doctor,* Hale & Iremonger, p.114
58. 'Australian Nurses in World War 1', https://ausww1nurses.weebly.com/
59. Harris, K. (2009). Red Reflections on the Sea: Australian Army Nurses Serving at Sea in World War 1. The Journal of Australian Naval History, 6(2), 51–73.https://search-informit-org.ezproxy.lib.uts.edu.au/doi/10.3316/informit.261756961849214, p. 53-54

The contingent disembarked in Egypt, to avoid overcrowded military camps in England over winter.[60] The entry of Turkey into the war on 12th October made this location even more strategic. An Australian camp was established at Mena, a short distance from Cairo and close by the Giza pyramids. 25,000 troops were camped there at its height, training six days a week. The nearby Mena House became a military hospital, not just for Australians but for other British forces stationed there.

The site of the AIF camp at Mena, adjacent to the pyramids
State Library of South Australia, Reference number B 61384.

The *Kyarra,* in its turn, arrived in Alexandria in January 1915, carrying the necessary equipment for Nos. 1 and 2 AGH and the remaining nursing staff of both.

No. 1 AGH was placed at the Heliopolis Palace Hotel at Heliopolis, while No. 2 AGH was initially located at Mena House Hospital, adjacent to Mena camp, where the majority of the Australian forces were located. When the number of wounded increased, it was later extended to the Ghezireh Palace Hotel, with the Mena House Hospital absorbing patient overflow. No. 1 ASH was stationed at the military camp at Maadi, while No. 2 ASH was to go in to camp at Mena, specifically for the treatment of venereal diseases. No. 1 ACCS was first placed at Heliopolis, then sent to Port Said.[61]

The Heliopolis Palace Hotel was an immense structure but initially, most of it was used for accommodation and only the Rotunda and Great Hall

60. Beard, D. (2022) Kangaroos by the Pyramids, AWM: https://www.awm.gov.au/wartime/98/articletwo
61. Barrett and Deane, The Australian Army Medical Corps in Egypt, p. 22

were used for patients.[62] When patients overflowed from the smaller Mena House Hospital, they were sent to the Heliopolis Palace.

Two additional camp hospitals were established at Heliopolis, one for venereal disease and one for infectious disease. The venereal disease cases were later moved into sections of the Heliopolis Palace, before a further move to an Abbassia barracks. The scope of No. 1 AGH was extended to Cairo's Luna Park, while another area was set up at the Casino for patients suffering from infectious diseases.

The hospital at Mena House Hotel, Mena, Egypt, 1915. AWM, J02564.

When the Gallipoli campaign started, 16,000 wounded men were treated in Egypt in the first 10 days[63] and a further convalescent hospital was set up at Al Hayat, Helouan.[64] When numbers of wounded increased still further, another hospital was set up at the Atelier, formerly a joinery factory, followed by another at the Sporting Club, near the Heliopolis Hotel. Later a convalescent wing of No. 1 AGH was established at the Ras el Tin School at Alexandria. But with accommodation still insufficient, the Grand Hotel at Helouan was converted into a convalescent hospital for Australians, before converting to a British Hospital. The Montaza Hospital at Alexandria also became a convalescent hospital for use by both Australian and British troops.

Such is the context in which Louisa wrote her diary.[65]

---

62.   Barrett and Deane, The Australian Army Medical Corps in Egypt, p. 22
63.   ibid, p. 35
64.   ibid, p. 36
65.   The diary lacks punctuation in many instances so basic punctuation has sometimes been added to aid the readability of the text

*A note on style: In her diary, Louisa used different styles for recording dates, sometimes including day of the week in full or abbreviated form; sometimes omitting month and year, especially as the diary progressed. Occasionally she grouped several dates or repeated a date at the end of an entry. In general, these have been standardized to one style to enhance clarity.

In both diary and letters, she frequently used abbreviations and where these may cause confusion, the full word is given in brackets. She often used the ampersand (&) but these have been changed to 'and', again for readability. She often omitted punctuation – for example, omitting full stops and apostrophes - or used then current styles, and where this does not interfere with meaning, such characteristics have been retained. In the diary, she uses a symbol - č - to indicate 'with' and these have been kept, with the word added in brackets.

She used a variety of styles to indicate 'as above' and sometimes left a dash or no entry after the date. Illegible words occur and appear in italics and doubtful transcriptions are followed by (sic?). The spelling of place names has changed in some instances but, in general, her original spelling has been kept in this transcription. Where possible, people and places mentioned have been identified and brief details given in footnotes. In some cases, however, this has not been possible.

Louisa's letters, reported mostly in the *Maitland Daily Mercury* and mentioned in this text have not all been given in full. Also, longer letters have been given in paragraph-form to assist readability and some punctuation has been added at times for the same reason.

Louisa's Diary Transcription

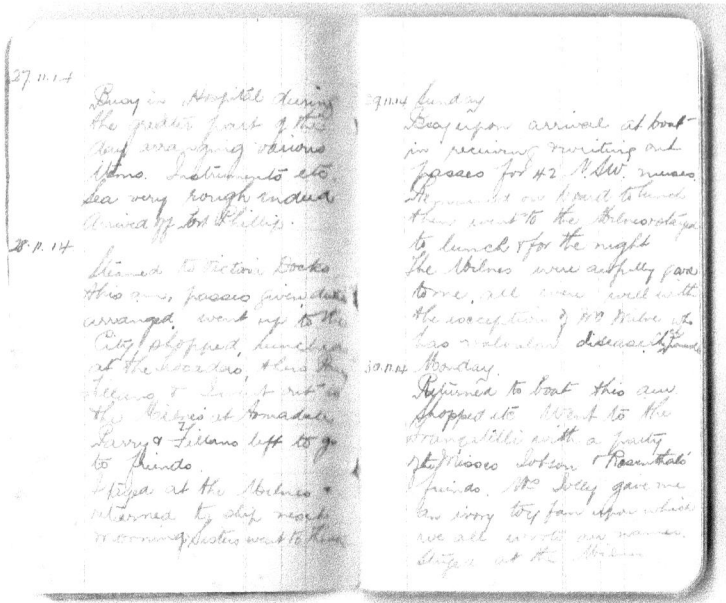

26.11.1914   **November 1914**

An accident occurred at 3.30 a.m. The ship's quartermaster's head was badly cut. De Quist (sailor) received a fractured leg with injuries to his knee and was operated upon at 11.30 a.m. Great push to get things from their various cabins for the operation as they were only shipped at Sydney.[66]

Was busy until after lunch in Hospital then feeling queer went to cabin which is being shared by Sisters Fillans[67] and Parry[68] and remained there.

27.11.14

Busy in Hospital during the greater part of the day arranging various items. Instruments etc. Sea very rough indeed. Arrived off Port Phillip.

28.11.14

Steamed to Victorian Docks this a.m., passes given, duties arranged, went up to the City, shopped, lunched at the Trocadero,[69] then Parry, Fillans and I

---

66. Kirsty Harris notes that operations and surgical treatment were performed on board ships in the first convoy and this entry indicates that the same was true for the *Kyarra*.

67. Wilhelmina Fillans was from Arncliffe, NSW. As the diary progresses, Louisa refers to her as 'Fill'

68. Annie Bertha Parry was from Picton, NSW. Sister Parry was notable for her later work with the St. John's Ambulance Association in New South Wales

69. The Trocadero was a theatre in Melbourne in this era but, from Louisa's description, it may have had a café attached

went to the Milnes' at Armadale.[70] Parry and Fillans left to go to friends. Stayed at the Milnes and returned to ship next morning. Sisters went to theatre.

29.11.14 Sunday

Busy upon arrival at boat in receiving and writing out passes for 42 N.S.W. nurses. Remained on board to lunch, then went to the Milnes and stayed to lunch and for the night. The Milnes were awfully good to me. All were well, with the exception of Mr. Milne who has valvular disease *illegible*.

30.11.1914 Monday

Returned to boat this a.m. Shopped etc. Went to the Francatelli[71] with a party of the Misses Jobson and Rosenthal's friends. Mrs Jolly gave me an ivory toy fan upon which we all wrote our names. Stayed at the Milnes.

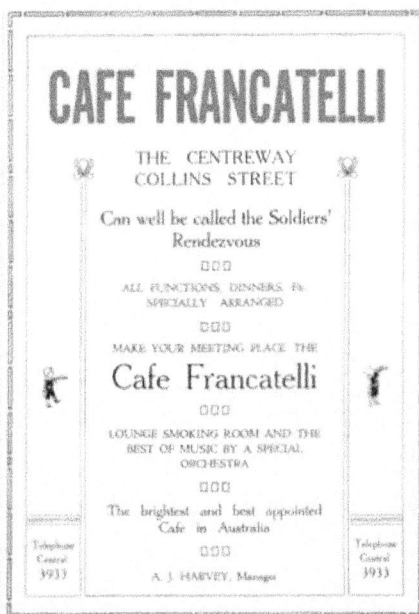

Louisa visited Café Francatelli during the *Kyarra*'s transit through Melbourne.
National Library of Australia, nla.obj-62637998.

---

70. Ada Milne and her family were cousins of Louisa's
71. The Francatelli was a café, located in The Centreway, Collins Street, Melbourne.

December 1914

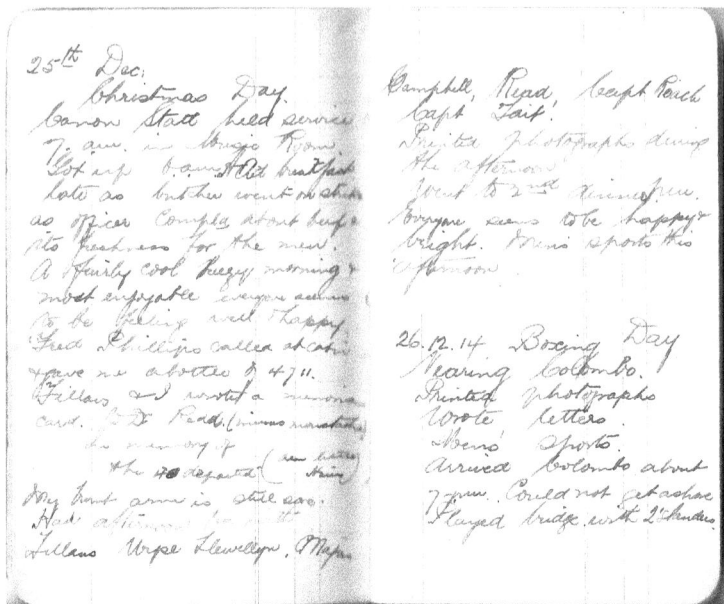

1.12.14 Tuesday

Returned to boat, had lunch at Mutual Stores[72] with Fillans and Mrs Lay. Afterwards went to "Spook" in Block Arcade.[73] Met Mrs Creaton (sic?) Met Capt. Reiach[74] in Perdriaus.[75] Slept on boat. Went to Wongonella, then to Armadale.

2.12.14

Wed: Relieved by Matron Knowles.[76] Went to *Kyarra* a.m. to report. Wrote out lists re inoculations. Stayed aboard to lunch. Bob Stobo[77] came to *Kyarra* and took Sister Spalding,[78] Stone[79] and I over printing works. Went to Armadale. Very tired.

---

72. The Mutual Store was Melbourne's first department store, located on the corner of Flinders and Degraves Streets
73. The Block Arcade, built in 1892, connects Collins Street and Elizabeth Street, Melbourne
74. James Reiach, Captain, attached to No. 2 AGH
75. Perdriau Rubber Company, Elizabeth Street, Melbourne
76. Mary Jane Knowles was a South Australian nursing sister
77. Robert Spears Stobo, 4th son of Edye Stobo. Robert Stobo was a Lieutenant in the Royal Australian Navy
78. Florence Ethel Spalding was from Manly, NSW
79. Constance Adel Stone was from Orange, NSW

3.12.14 Thursday

Reported at *Kyarra* and Fillans and I saw the Spears[80] and stayed with the Glencross.

4.12.14 Friday

Returned to boat 10 a.m. Bob Stobo took us out to St Kilda thence to Capt. Wills, Scott St. East, St. Kilda for the morning. Myself and Fillans went over through Prahran, to dentist then to Armadale.

After leaving Capt. Wills, where we met Miss Wills, Mrs Downie who sang beautifully and Miss Johnson (pianist), we missed last tram 11.02 p.m. and had to motor to Armadale. Drive was delightfully cool. Ada and Maggie both came out to me upon my return.

5.12.14 Saturday

Met Bob Stobo and Parry at Flinders St Stn. We went to the Mia Mia[81] for morning tea. Bob presented us with a 2 lb. box of chocolates each (non *illegible* on return trip) we took a cab returned to the ship at 12 midday. Fillans arrived likewise. All sisters detailed to wharf during afternoon. We sailed from the docks about 6 p.m. Bob and 2nd officer with a friend came down to see us off. We got some snapshots before leaving.

6.12.14 Sunday

All feeling very queer and stayed in our bunks all day. No Good! Wretchedly rough day č (with) swerves (sic?).

7.12.14 Monday

Arose! Managed to come on deck about 4.30. Retired about 8.30 p.m.

8.12.14 Tuesday

Was on duty 6-10 a.m. Feeling squeamish, 1st Hosp. supposed to be "on duty" "pig. dwags".[82] Felt better after a little exercise. Stayed on deck all day, had a game of bridge[83] with Parry,

---

80. Arthur Spears and his family, who lived in Brunswick, were cousins of Louisa's
81. Mia Mia Tea Rooms were at 340 Collins Street, Melbourne
82. Possibly sausages or frankfurts wrapped in pastry
83. Louisa's family were inveterate bridge players

Martin[84] and de Boissier.[85] Non comms. (non-commissioned officers) gave an enjoyable concert 8 p.m.

9.12.14 Wednesday

Got up for breakfast, went on deck, was inoculated for typhoid 250.000.000 and 100 ggs. (sic?). Stayed on deck. Weather calm and bright. Wyse[86] showed me on upper deck how to fix photographs. Went to cabin to rest, had dinner downstairs, got up for the evening, played bridge with Martin, Garven[87] and Freeman (sic?). Had supper, retired, am reading "Golden Deeds" by Charlotte Yonge,[88] very enjoyable and interesting. Had a little French[89] with Llewellyn[90] and Richmond.[91]

10.12.14 Thursday

Arose early and went on deck before breakfast. Wrote up diary and a few letters.

Was up on deck with Wyse, printing photographs. Read "Golden Deeds" etc. Watched deck hockey and billiards being played and dancing fun. Had French č (with) Col. Springthorpe[92] and others.

11.12.14

Got up early, went on deck, wrote letters, had a French class with Col. Springthorpe and Capt. Mattai,[93] this to be a daily affair. Day very calm and cloudy. Played auction bridge with Sisters Morrice,[94] Holloway[95] and *illegible* and won hands down. Arrived Fremantle 9 a.m.

84. Mary Theresa Martin was from Newcastle, NSW
85. Phyllis Mary Boissier was from Mascot, NSW, and held the ranks of head sister and temporary matron. The AWM refers to her as Phyllis Boissier
86. Myra Septima Wyse was from Randwick, NSW
87. Ida Mary Garven was from Clarence River, NSW
88. Yonge, Charlotte M., A Book of Golden Deeds of all times and all lands, London, Macmillan, 1864
89. French lessons were given on board the *Kyarra*. John Nash, Lieutenant Colonel, who was also on board, recorded the large attendance of doctors and nurses at these classes, conducted by a former teacher at St Aloysius, North Sydney
90. Gwladys Gwynne Llewellyn was from Ryde, NSW
91. Daisy Donaldson Richmond was from West Goulburn, NSW
92. John William Springthorpe, Lieutenant Colonel, AAMC; Senior physician with No. 2 AGH
93. Captain Charles Mattai, physician
94. Nellie Constance Morrice was from Sutton Forrest, NSW
95. Eva Estelle Holloway was from Ashfield, NSW

12.12.14

Kept awake a good deal with coaling of ship and noise of electric fan. Ports were closed which made the air unpleasant. Expected to go ashore this a.m., disappointed, beautiful day. Received letters from Mary and Lilly Simpson č (with) Aggie Crawford's address, "The Shieling", Maybole, Scotland therein. Went ashore before lunch. A lot of telegrams to mail: and to Ede,[96] posted letters. Had a cup of tea in Fremantle with Fillans and Wyse. Went to Perth with Wyse, went out on train to suburbs, walked from one tram terminus, Leederville to North Perth. Thence we went to Perth city to the Court House gardens,[97] saw several streets, had tea at the Savoy[98] and then went to Spencers Pictures,[99] returned to ship, found all ports closed. Fillans and Parry stayed ashore, saw Heriot (sic?).

13.12.14

Stayed aboard a.m. and went to Church and Parade in Fremantle. The archbishop of Perth preached. Canon Statt[100] assisted. Quite a large number of Sisters, Officers and men attended the Service. After lunch Llewellyn, Wyse and I went to Subiaco to see Mrs Ambrose (Nurse Young) who has a nice cottage home; her six children are lovely. We met a Lieut. Taylor there. Dr Ambrose was away in the country.

We then went into Perth by train to the Queen's Gardens[101] and finished up by having tea on the Esplanade and then made some investments at a chemists and returned by the 9.30 p.m. train to Fremantle. The twilight lasts till quite late. After having had tea we went along the avenue, saw Govt. House, the Cath. (Cathedral) etc.

14.12.14 Monday

Having had instructions to be aboard by 12 m.d. (midday) we went into Fremantle, made some purchases, had morning tea and returned to the boat. The tea aboard is not at all nice. We sailed from Perth at 7 p.m. but were unable to go ashore meanwhile. I wrote 3 letters while waiting. A fairly large no. (number) of people came to the wharf to see the ship off. Sutton (W.H.) was on the wharf.

96. Louisa's younger sister Edith
97. Court House Gardens, (Supreme Court Gardens) CBD, Perth
98. The Savoy Hotel, Hay Street, Perth
99. Spencers Pictures, Theatre Royal, Perth
100. Gordon William Statt, Staff Sergeant, Church of England.
101. Queen's Gardens, a park in East Perth.

15.12.14 Tuesd.

Sea was fairly rough but we were not seasick and have been able to go to the saloon for meals almost even since entering the G. A. Bight (Great Australian Bight). Hemmed a tray cloth today. Had a game of hockey.

16.12.14 Wednesday

Had another injection of typhoid vaccine. 500.000.000 which seemed to have a more severe effect than the last dose. I felt very drowsy during the afternoon and had a sleep. Played cards with Capt. Alcorn[102] and Sabine[103] and Miss Campbell.[104] 2 stand. Worked names on the cloth.

17.12.14 Thursday

Went by request with Miss Bell to the C.O.'s office[105] to receive instructions with regard to the Nurses and was kept busy almost the whole morning.

Filled in some more autographs. Was very tired and stiff so went to cabin to rest p.m. Ships alarm sounded twice during the afternoon. Everyone had to go to their boat stations and again a second time, several girls stayed below to dress themselves. Played cards p.m. with Capt. Sabine, Sisters Campbell,[106] Foster[107] and Scully.[108] Retired 10 p.m.

18.12.14 Friday

Not well this a.m. Got up for 1st Break (breakfast); walked a mile after breakfast, worked autograph cloth.

19.12.14 Sat.

Men's sports p.m. Very hot day, weather continuous fine and calm.

20.12.14 Sun.

Sighted Cocos Islands. Church parade.

21.12.14 Mond.

Sisters and Drs (Doctors) sports cont. (continued) p.m. Finished reading Golden Deeds (Charlotte Yonge very interesting).

102. Alfred Alcorn, born 1892 East Maitland NSW
103. William Sabine, Captain, was attached to the No. 1 AGH
104. Beryl Campbell, Matron was from Rockhampton Queensland. She was attached to No. 1 AGH
105. It could be inferred from this that Louisa occupied one of the senior positions among the nurses on board
106. Lily Campbell was from Richmond River, NSW
107. Edith Annie Foster was from Vaucluse, NSW
108. Mary Scully (aka May) was from Dalby, Queensland and was later Matron of both the Enoggera Hospital and the Rosemount Hospital, Brisbane

22.12.14 Tuesd.

Officers sports and pillow fight over greasy pole above a tank of water. Very humorous affair. Read 2 books lent by Sister Boissier. Ironing č (with) electric iron, cabin, burnt hand.

23.12.14 Wed.

Perspiration simply dripping from face. Got up very early 5 a.m. Sighted strange steamer and vessel changed its course, all sorts of rumours going round and bound for Africa etc. etc. Stewards = iron crossing (sic?) smoke etc. and crossing the line. Inoculated for the 3rd time 100.000.000.

24.12.14 Thurs.

Arms painful. Got up soon after 5 a.m. Nights and days very steamy. Cabin close. Worked autogs. (autographs) and read. Neptune came aboard p.m. Played bridge with Capts. Lee, Sabine and Alcorn, and Sister Campbell. Stewards sang carols 12. *Illegible*.

25.12.14 Dec. Christmas Day

Canon Statt held service 7 a.m. in Music Room. Got up 6 a.m. Had breakfast late as butcher went on strike as officer complain'd about beef and its freshness for the men.[109] A fairly cool breezy morning and most enjoyable, everyone seems to be feeling well and happy. Fred Phillips[110] called at cabin and gave me a bottle of 4711.[111] Fillans and I wrote a memoriam card to Dr. Read[112] (minus moustache) in memory of the 40 departed (dear little hairs). My burnt arm is still sore. Had afternoon tea with Fillans, Wyse, Llewellyn and Majors Campbell, Read, Capt. Reiach, Capt. Tait.[113] Printed photographs during the afternoon. Went to second dining p.m. Everyone seems to be happy and bright. Men's sports this afternoon.

109. Walter Rainsford, who was an orderly room clerk with the No. 2 AGH, noted in his diary that there was 'plenty of tucker' on board the *Kyarra*, but that the meat was sometimes bad. Barrett and Deane's history of the AAMC records that 22 officers and others were infected with ptomaine poisoning, with two serious cases eventuating

110. Frederick Stobo Phillips was Louisa's nephew by marriage. He enlisted in the war as a Private in the Medical Corps. He was awarded the Military Medal for conspicuous bravery at Pozieres as a stretcher bearer, and was given a commission as Second Lieutenant with the 1st Battalion. He was killed in action on the Somme on 5 November 1916

111. A popular cologne, reputed to be cooling in hot weather

112. William Henry Read, Major, was attached to No.2 AGH. He was married to Irene Phillips, sister of Frederick Stobo Phillips, both of whom were related by marriage to Louisa

113. John Thomson Tait, Captain, was attached to No. 1 AGH

26.12.14 Boxing Day

Nearing Colombo

Printed photographs

Wrote letters

Men's sports

Arrived Colombo about 7 p.m. Could not get ashore. Played bridge with 2 standers.

27.12.14

Went ashore in a pulling boat about 7.30 a.m. went direct to Railway Station in a ghari.[114]

The foreign sights, sounds etc. interested us immensely, native carts, rickshaws, barbers shops, beggars etc.

We were too late for the train to Kandy so took a train up to the Business Quarter. We went to a Hindoo's store and invested in a few odds and ends, the various ornaments, silk kimonos (sic?) etc. are in many instances rare and beautiful. We then returned by train to the Railway Station and went by a special train (11.30 a.m.) to Kandy. The trip was much enjoyed and the scenery is wonderful especially when going up from the valley to the heights, a wonderful rise of 2000 feet in 1 hour. The Bible Rock[115] which was a refuge for natives during a siege. This line is said to be one of the greatest engineering feats in the world. This engineer who built the line was honoured by having a large monument erected at the upper end of the line.

All along the railway line from Colombo to Kandy the tropical growth consists mainly of coconut palms which had large clusters of nuts on them. (We bought a King Coconut and drank the contents which seemed to taste like banana water. We also bought small plantain bananas, very sweet). The growth also consists of plantains, lantana (scarlet) numerous trees, foliage, vines etc. and every now and again pretty creeks are reach(ed) also swampy marshes which are used to a great extent for the cultivation of rice.

The slopes of the hills are in many instances terraced in continuous small terraces with a low retaining wall of mud. This acts as a dam and the rice is

114. Gharry, a horse drawn vehicle
115. Bathlegala was known as 'The Bible Rock' because it was said to resemble an open book and by association how Mount Sinai might look. It became a place of refuge in the conflict between the Sinhalese with Portugal, allowing for an early warning system

grown within. The native huts and villages are very picturesque but anything but sweet (a peculiar odor always being present in native quarters). The walls of the huts are mainly built of mud but sometimes of a poor brick and are in some cases, coloured fawn, white, dull red etc. etc. a fair number seemed to be built in ordinary shed style with a division midway across the door of the room, opened into the opposite half which had no front wall and thus were built in one long row and were divided into very small rooms which apparently were occupied by various families. The roofs of these houses were either covered by small, ridged tiles, red or were thatched with woven palm leaves which apparently prove to be quite weatherproof.

All along the line at the various stations natives brought bananas, quinces, plantains etc. for sale and endeavoured without exception to obtain thrice the value of their goods. At a station above the monument a large crowd of villagers came to the train, old and young. Snapped the crowd and hope the result is good. The youngsters were very eager to get coins given them and in one instance, tried to make an old, very thin woman, beg.

We arrived in Kandy early in the afternoon, went direct to the Queen's Hotel[116] which is very good indeed. £1. Kandy is simply a series of pictures and we are almost impelled to keep on humming "From Greenlands icy mtns" (mountains)[117] for this hymn truly describes Ceylon.

The roads are well kept and with limbs of clean barked trees overreaching them. The green foliage, the numerous native carts, the natives (old and young) the coloured oriental costumes, the various buildings, native shops and quarters, the temples, flowering trees and shrubs etc. (It is beautiful).

The hotel rooms were large (take 4 single Metropole rooms therein) with polished floors, oak furniture (dressing tables, washstands, wardrobe etc.) with four poster beds draped with mosquito nets. The rooms are very well ventilated and open into wide hallways and are fitted up in each instance with electric light and large electric fans.

We had afternoon tea brought upstairs then sallied forth to see all we could. The lake is perfect and was artificially formed during the reign of one of the Kandian Kings.

There is an island therein upon which the ladies of the King's harem were

116. Queen's Hotel is one of Sri Lanka's historic hotels, located at DS Senanayake Veediya, Kandy
117. From Greenland's Icy Mountains by Reginald Heber

wont to take their recreation. They gained access to the island by means of a subterranean passage from their quarters. "The Temple of the Tooth"[118] was approached in the same manner from the King's Palace.

The Queen's Hotel, Kandy, AWM, CO3709.

The Bund is bordered on the lakeside by huge bosses of masonry which gives a very picturesque air to the place. An equestrian statue was erected in a square just east of the hotel in memory of the soldiers who fell during the S.A. (South African) War.

After leaving the hotel we were immediately besieged with rickshaw men but instead of employing them we entered some native shops and made some small investments. Later we wished to visit the Temple but a native was so persistent that we decided to shake him off by continuing our tour of inspection, reached Govt. House grounds, the entrance being guarded by a native policeman who did not speak English.

We saw a gentleman coming down the slope, made enquiries of him, he proved to be a Mr. Aakins who has a coconut plantation and desiccated coconut plant near Colombo. He kindly offered to show us the way and walked up to the top of the hill (Lady Horton's Drive) the view of the Lake hills and surrounding countryside being very fine. The colouring is beautiful.

118. Sri Daladwa Maligawa, Temple of the Sacred Tooth Relic is a Buddhist temple and part of the Royal Palace Complex. It is said to contain a relic of the Buddha's tooth

Mr. Aakins then escorted us to the Temple of the Tooth. The Temple was shut but we saw in an octagon shaped building the old books etc. in many instances being hundreds of years old. The Buddhist priests wear yellow draperies and have shaven faces and closely cropped heads. They only have 1 meal a day (breakfast) 1 cup of tea in the evenings, yet they look well and fit. They are not allowed to touch money but are kept by donations in kind.

We were fortunate enough in being allowed to go through a trap door into a cellar beneath the octagonal building and saw the priests sitting room (quite big) with an old passage leading to the King's Palace which is now locked up, also the windows which are approached by means of narrow straight stairs with a platform above, upon which the Queens were allowed to see the Autumn Procession.

We were upset by the sight of maimed blind people asking for alms, one man had no legs and was lying on his back. As people passed, he 'worried' himself around and held his grimy looking stumps up to view. Next day he was standing upon his stumps. Little children escort the blind and beg for them. "Ladee! Ladee!" and saluting us in a deprecating manner. One gets quite tired of the continual begging of the people all round though the higher class appears to be rather a sedate educated person, many as a mark of distinction wear tortoiseshell combs in their hair (semi-circular).

We returned to the Hotel to dinner, escorted to the Hotel by Mr. Aakins, and afterwards went for a short walk. Retired early.

28. 12. 14

Arose early 5.45 a.m. in order to go to Peradenyia Gardens,[119] had morning tea in our rooms and drove in a ghari to the Gardens which are said to be the largest in the world. They are certainly all beautiful, all sorts of choice flowers, foliage with great clumps of bamboo. We saw and gathered cinnamon, cloves, nutmeg, nuts etc. The natives were making their toilets in the open in front of their homes in small 1 pint basins. Some were washing their faces, combing hair and cleaning their teeth. We wrote our names in the gate book. We reached the hotel for breakfast, then Sister[120] and I went to the temple, saw the casket containing the tooth which an eyewitness says

119. Royal Botanic Gardens, Peradeniya are 5.5 km west of Kandy
120. Possibly Daisy Richmond

31

must be several inches long. Afterwards we went for a walk around the lake incidentally seeing the 2nd Buddhist temple. We met Fillans and Wyse and took photographs of one another in a rickshaw, met illegible and others. The priest at the 2nd temple allowed us to photograph himself and a little student standing beside the tomb of Buddha's bowl.

Returned to hotel, bathed and dressed, then lunched and went for a rickshaw ride to see the elephants at Katugastota.[121] Four animals appeared. On the return journey we entered a Mohammedan Temple which was dingy and poor. One man was lying asleep on the floor. A Mohammedan priest (a Turk) walked part of the way with us and amongst other things said if this present European war was a religious war, all Mohammedans would have to join in.

After returning to Kandy we visited some shops and then walked to the Railway Station, reaching Colombo at 9.50 p.m. The fireflies were quite numerous, the night was moonlit and at the various stations, fruit sellers etc. were in evidence. At one place a native procession of some sort was going on.

At Colombo we took a tram but when about ½ way to our destination it turned off to the sheds and we had to walk the remainder of the distance to the boat wharf (from R.C. Church). We came in a pulling boat to the ship with Crago,[122] Llewellyn and others. We had fruit and lemon squash, retired, night fairly warm, slept on deck.

29.12.14

Breakfasted aboard, went ashore in pulling boat went for motor to Mt. (Mount) Lavinia[123] Cinnamon Gardens[124] etc. shopped, called at Govt. House to see Mrs Frazer (Mrs J.A. Dean's daughter). Had lunch with Frazers and a friend, Mrs Harding, returned to ship, was given heather and mistletoe by Scotchman at Cargills[125] Co'bo (Colombo).

Played cards 2 standers. Ship sailed. Night very warm.

---

121. The Katugastota elephant bath was a popular tourist attraction, located some 4 km from the Kandy city centre
122. Bessie Crago was from Yass NSW
123. Mount Lavinia is a suburb of Colombo, located near the 'Golden Mile of Beaches'
124. Cinnamon Gardens is a suburb 3 km south-east of Colombo's centre
125. Cargills was a department store and providore in Colombo

30.12.14

Got up fairly early, read "A Knight on Wheels".[126] The day was terribly warm and stormy. Went to cabin despite the heat and slept until after lunch. Parry brought me some fruit.

Read on deck during the afternoon and played cards č (with) Sisters Campbell (2) and Scully. Reading on deck p.m. Slept on deck but was uncomfortable.

31.12.14

Got up very early as the upper deck was being washed down, had a cold and lengthy saltwater bath, rested in cabin until 7.40 a.m. Arranged list re Nurses' duty, med (medical): or surg. (surgery). Finished reading "A Knight on Wheels".

A fancy dress masked ball was held on the port deck this evening. The costumes were remarkable for their variety and effectiveness. I went as a daughter of the regiment.[127]

---

126. Hay, Ian, A Knight on Wheels, Hodder and Stoughton, 1914
127. 'A Daughter of the Regiment' was a comic opera by Gaetano Donizetti

January 1915

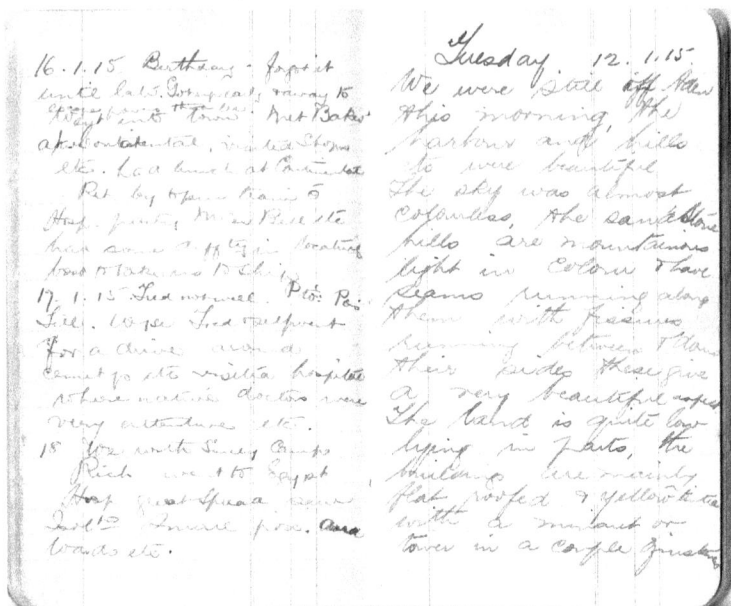

Entries for 16-18 January 1915 and second entry for 12.1.1915 [128]

1.1.1915

New Year's Day. Nothing remarkable happened. Spent the day reading etc. Was very drowsy after keeping late hours yest. (yesterday). And early rising this a.m. too much noise on deck and then the men commenced washing the upper deck which is an effective means of dispersing of the sleepers. Printed a number of photographs (sic?).

2.1.1915 Saturday

Mens sports.

3.1.15

Got up late. Went to Church Service. Singing on deck during the evening service conducted by Col. Nye.[129] Some of the AMC[130] (Army Medical Corps) attended and the service went off very well indeed.

---

128. This second entry has been placed after the corresponding date in January 1915
129. Edward Nye was a Methodist Minister of Religion. John Nash recalled that he was a Baptist Methodist from Victoria, who, while holding no substantive rank, was entitled to be saluted and addressed as Colonel
130. Australian Army Medical Corps

4.1.1915

Drs (doctors) sports. Final heats. Printed a large budget of photographs.

6.1.15 Tuesday

Got up about 6.30 a.m. Fillans feeling better this a.m. The African coast was to be seen on our left. Cape Guardafui[131] and the Apostles' Island.[132] The weather is still clear calm and cool. Fixed all printed photos done yesterday. Passed Barren Rock Island.[133] Arrived at Aden[134] p.m. The mountainous headlands are very fine and rise up with steep cliffs in places and show against the sky with sharp serrated edges rising into sharp peaks.

The right hand side upon entry was dotted with numbers of windmills and there appears to be a tiny beach about midway along the face of the cliffs. We could plainly see the water tanks etc. The sea was a very pretty green and the heights when in shadow are quite a dark smoky grey which, where the sun strikes them, turns into a semi-golden brown.

The Cruiser, *Empress of Russia*,[135] was lying in port and patrols the Red Sea. There were several other ships in port and the sailing and other boats made a most picturesque effect. Numbers of small boats came alongside with yelling crews of natives aboard, selling sweets, fans etc. They make one tired of their continuous noise. They throw a weighted cord up to the deck to which is attached a small basket in which they place any purchases of the passengers.

I did some ironing, had a very warm time of it. The ship sailed at 10.30 p.m. *The Empress of Russia* left anchorage early in the evening. The Mountains and Bay looked very fine in the moonlight; the moon being somewhat hidden by clouds. Col. (O.C.)[136] went ashore for orders.

8.1.1915

The Empress of Russia and armoured merchantman "*Himalaya*"[137](8 guns) are patrolling the Red Sea. The latter held up the German prize S.S. *Freienfels* [138]

131. Cape Guardafui forms the apex of the Horn of Africa
132. Possibly one of the four islands and two rocky inlets which are located off the coast of the Horn of Africa
133. Possibly the Guano Islands of southwestern Africa
134. Now known as Yemen
135. *The Empress of Russia* was an ocean liner requisitioned by the British Admiralty in World War 1. After the Battle of Cocos, survivors of the sinking of *The Emden* were transferred by HMAS *Sydney* to *The Empress of Russia*
136. Walter Rainsford also notes in his diary that Colonel Martin and a small group were the only ones who went ashore at this point
137. The *Himalaya* was a Royal Navy Merchant Cruiser
138. The S.S. *Freienfels* was a German-built cargo steamship, seized by the United Kingdom in 1914

manned by a prize crew.

We sighted 3 ships on our portside and later the *Himalaya, Empress of Russia* and S.S. *Freienfels.*

We saw the Island of Perim[139] at this time which was taken from the French about 100 years ago. One tale says that someone ?[140] arriving from the Red Sea stated that all was clear, the French guard went out to sea, when they returned the British flag was flying.

Perim is the key to the Red Sea, without it Aden would be useless to Britain. Perim is at the Nth end of the Strait of Babel, Babel Mandeb (Bab-el-Mandeb).[141]

The Scottish-built *Empress of Russia,* a steam turbine liner, was refitted as an armed merchant cruiser in World War 1.

We later on sighted Mocha[142] which is on the Starboard side of the ship. We could see the lighthouse, the flat roofed buildings, an obelisk standing out about the middle of the buildings. The view was hazy due to mist and some say dust.

We have a cool breeze blowing today.

We again sighted land later on in the day and passed Aboukir Lighthouse[143] which was on our Starboard side. The small patrol ship was within touch with us during today. *Freienfel.* Played bridge with Col. Nye and others.

139. Perim Island was located in the south entrance to the Red Sea
140. This is Louisa's own question mark
141. The narrow strait that separates Yemen, on the African Peninsula, and Djibouti and Eritrea, in the Horn of Africa
142. Now spelled Mokha, Yemen's main port
143. Abū Qīr

9.1.15 Saturday

Still steaming along the Red Sea. The patrol ship was in sight during night. Some of the sisters sleeping on deck were quite excited during the night thinking they were being followed also that the moon was an aeroplane.

Today we have somewhat of a swell on, a ship meeting us off the Starboard side was dipping at a great rate.

The day has been very warm, washed h.[144] Finished reading Miranda on the Balcony[145] and commenced to read "The Amateur Gentleman"[146] by Jeffrey Farnol. Church a.m.

11.1.15[147]

Entered G. of Suez (Gulf of Suez).

Illegible 7 a.m.

Israelites supposed to have crossed at many spots.[148]

12.1.15 Tuesday

Natives on board and alongside selling various goods. The early morning sights were fine. Land low lying along the canal. Indian troops and others guarding shores.

12.1.15 Tuesday[149]

We were still off Aden this morning. The harbor and hills etc. were beautiful, the sky almost colourless, the sandstone hills are mountainous, light in colour and have seams running along them with fissures running between and down their sides. These give a very beautiful aspect. The land is quite low lying in parts, the buildings are mainly flat roofed and yellow tinted with a minaret or tower in a couple of instances. The air is quite chilly and the seagulls are flying about in large numbers and quarrelling over food, squawking and crying.

144. Possibly washed hair
145. Mason, A.E.W., Miranda on the Balcony, London, Macmillan 1899. It was said to be a modern retelling of Homer's Odyssey, and a reference source for James Joyce when he wrote Ulysses
146. Farnell, J., The Amateur Gentleman, London, Sampson, Low, Markson & Co, 1913. The novel traces the psychological development of its main character,
147. There is no entry for 10.1.1915 but two entries for 12.1.1915, suggesting that Louisa may have briefly confused the dates
148. Louisa is referring to the Biblical story of Moses and the Israelites crossing the Red Sea
149. This second entry for Tuesday 12 January appears several pages later in the diary. It may be that Louisa skipped some pages and then went back to use them. The entry has been placed after the initial entry for the purpose of maintaining the chronology

Numbers of boatmen (native), with fairly large sailing boats with one sail, came alongside selling fruit (oranges, apples, dates and figs) sweets, cards, beads etc. They are great frauds and try in every instance to overcharge for their wares. No mail received but one went off the ship. We left for the canal between 12 and 1 p.m. Scenery is very wonderful.

The swampy shining marshes in the foreground reflected surrounding buildings trees etc. (mainly palms) and was multicoloured. One obtained occasional glimpses of deep blue sea and now and again through the trees and buildings, this was just after entering the canal.

Local traders in Aden, ferrying goods to troops on board an Australian transport, via a set of ropes. AWM, CO2540

There is a naval station at the entrance, a large yellow building.

Trees are planted upon the left hand side of the canal; they are very dusty looking in appearance and we saw tables and chairs arranged beneath them in a couple of places near the edge of the canal.

There were several wharfs jutting out into the canal and many natives and occasional Europeans were seen and we passed many small sailing boats etc.

The canal is bordered on either side by earthworks, sandy in colour which are said to have been raised since the declaration of war. These are protected by large numbers of native troops (Indians I should say) many having turbans, which are placed at intervals of every few hundred yards. They hurrahed with a good turn upon the r. (return?).

13.1.1915

Around Pt Said (Port Said), wakened by natives coaling. We were allowed ashore, strolled around town sightseeing, had lunch etc. Drove this native quarter sin, sand (sic?), sun and flies, filthy houses and natives, visited various shops, had fortunes told, etc.

Quaint domed churches.

Sailed p.m. De Lesseps statue.[150]

Entered Alexandria harbor after a calm trip, had to remain ashore all day. Col. Martin had gone ahead, overland from Pt Said (Port Said), all awaiting his return anxiously.

15.1.1915 Friday

Decided to go to Cairo, wired Bob[151] from Stn (station), missed train owing to stupidity of Arab. Thermos broke, much excited and interested over scenery, native villages, camels, donkeys etc.

Arrived Cairo 12 m.d. drove to Shepheards Hotel,[153] took train to Mena [154] Mrs. (sic?) on train. Saw Bob after visiting Mena Hosp.[155] Saw Miss Gould,[156] Johnston[157] Sister Bessie, Pyramids.

Bob took me to dinner at Continental[158] and I stayed with Sister *illegible.* Saw Kellett[159] and *Illegible.*

Robert Scobie, Louisa's elder brother, second-in-command of the 2nd Battalion[152]

150. Ferdinand de Lesseps was the French diplomat who built the Suez Canal
151. Louisa's older brother was Robert Scobie, final rank, Lieutenant Colonel, 2nd Australian Infantry Battalion
152. Collection of Dinah Toohey
153. Shepheard's Hotel was the leading hotel in Cairo, built in 1841, and also known as the 'Hotel des Anglais'
154. Mena Camp was the main AIF camp in Egypt, located c. 16 kilometres west of Cairo
155. The Hospital at Mena House, Mena. It had been requisitioned by Australian troops
156. McCarthy, P. M., Ellen Julia (Nellie) Gould, (1860-1941). Australian Dictionary of Biography, Volume 9, 1983. Gould was Matron of No. 2 AGH
157. Julia Bligh Johnston was born at Windsor NSW in 1861. She was a senior nurse at No. 2 AGH
158. The Grand Continental Hotel was frequented by officers stationed in camps near Cairo
159. Adelaide Kellett joined the AANS the same year as Louisa and was theatre sister with No. 2 AGH

16.1.15

Birthday, forgot it until later. Got up early and away to escape powers that be. Went into town. Met Baker[160] at the Continental, visited shops etc. had lunch at Continental. Returned by open train č (with) hospital party, Miss Bell etc. had some diffty (difficulty) in locating boat to take us to ship.

17.1.15

Fred[161] not well. *Illegible*: Fill. and Wyse. Fred and self went for a drive around cemeteries etc. Visited hospital where native doctors were very attentive etc.

18.1.1915

We with Scully, Camp. (Campbell?) Rich (Richmond?) went to Egypt Hosp,[162] great spread, saw Isolation smallpox wards etc.

The Grand Continental Hotel, Cairo. AWM, CO3473

19.1.15

Went into town with the idea of visiting San Stefano,[163] met Major Simmons who took us to wool exchange where the din was great. Gave old chap Australian 3d. (three pence?) was very pleased. Visited Ramlegh Hospital[164] in all 1100 beds (5-6 in each).

---

160. Possibly Maud Baker who was attached to the No.1 Australian General Hospital
161. Likely to be Fred Stobo Phillips
162. Ghezireh Palace Hospital, location of No.1 AGH. It was sometimes referred to as the Anglo-American Hospital and became the location for No. 2 AGH later in 1915
163. San Stefano was a neighbourhood of Alexandria
164. A 1915 text on hospitals in Egypt during World War 1 by H.T. Ferrar, states that Ramleh Casino and San Stefano Casino in Alexandria are likely the same hospital. It was intended for infectious diseases

20.1.1915

Train to Cairo in charge of Col Nash,[165] marshalled at other end and went by tram to Mena, were welcomed and shown our rooms, all very comfortable. We half expected to go into tents. No. 1 girls remained behind.

21.1.1915

On duty all day. No. 6 wards very nicely arranged.

22.1.1915

Aust. Mail delivered. Ward No. 6. Went to camp.

23.1.1915

On duty all day.

24.1.1915

– (dash possibly indicates 'as above')

25.1.1915

–

26.1.1915

Went to Cairo č (with) Col. Braund[166] and Bob and Wyse. Fillans came with us and went home early č (with) Col. We went to the Rursaal (sic?), visited Mehmet Ali Mosque.[167] Col. gave each of us a piece of alabaster.

27.1.1915

On duty all day.

28.1.1915

Went to camp and had dinner č (with) Bob, Col. (Braund) and Major Gordon [168] and had dinner at Continental.

29.1.1915

On duty.

---

165. John Brady Nash, Lieutenant-Colonel, 2nd AGH
166. George Braund, Lieutenant-Colonel was Robert Scobie's commanding officer and a serving member of the NSW Parliament. After a spirited stand at Gallipoli he was accidentally killed by friendly fire on 4th May, 1915
167. Now known as the Muhammad Ali Mosque and situated in the Citadel, Cairo
168. Charles George Gordon, Major, 2nd Battalion AIF. He was killed in action at Gallipoli on 25 April 1915

30.1.1915

Bought a black costume at Cicurels[169] in Cairo. Came home to dinner.

31.1.1915

On duty.

Lieutenant-Colonel George Braund, (front row centre) who raised and trained the 2nd Battalion, died at Gallipoli on 4 May 1915, seen here with Robert Scobie (front row, second left). Sandra Roberts Collection.

169. Les Grands Magasins Cicurels was a chain of elite department stores established by Moreno Circurel, who emigrated from Turkey in the 1880s. They sold ready-to-wear clothes, as well as other goods imported from Europe

February 1915

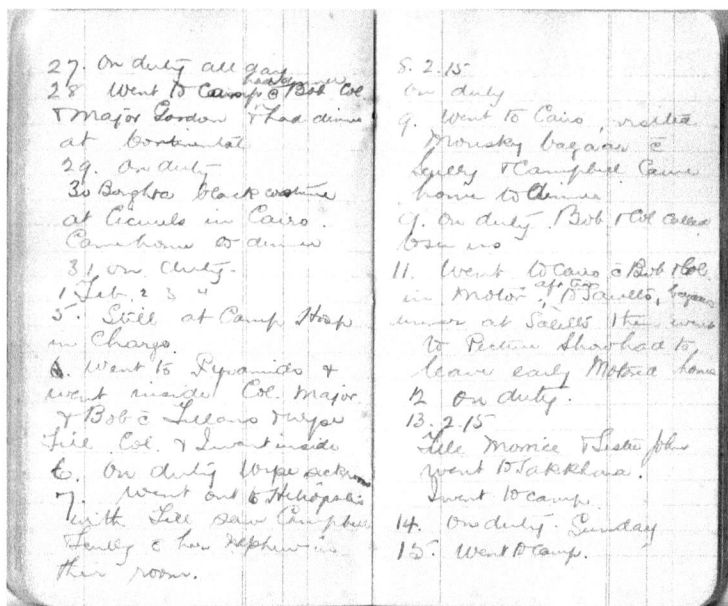

Entries for the closing page of January and the opening page of February 1915

1 Feb 2, 3, 4 - No entries for these dates.

5.2.1915

Still at Camp Hosp.[170] in charge.

Went to Pyramids and went inside Col. (Braund) Major (Gordon) and Bob č
(with) Fillans and Wyse. Fill, Col. and I went inside.

6.2.1915

On duty Wyse sickroom.

7.2.1915

Went out to Heliopolis with Fill, saw Campbell, Scully and her nephew in
their room.

8.2.15

On duty

---

170. According to 'Following the Twenty-Second,' 42 hospital marquees were pitched at Mena Camp,
35 for general patients and 7 for isolation cases. https://anzac-22nd-battalion.com/hospitals-egypt/

9.2.1915

Went to Cairo visited Mousky[171] bazaar č (with) Scully and Campbell. Came home to dinner.

10.2.1915

On duty Bob and Col. (Braund) called to see us.

11.2.1915

Went to Cairo č (with) Bob and Col. (Braund) in motor after tea to Sault's, bazaars, dinner at Sault's[172] then went to Picture Show,[173] had to leave early, motored home.

12.2.1915

On duty.

13.2.15

Fill, Morrice and Sister Johnst. (Johnston) went to Sakkara.[174]
I went to camp.

14.2.1915

On duty Sunday.

15.2.1915

Went to camp.

16.2.1915

Mjr (Major Gordon), Col. (Braund) and Bob called. On duty all day.

17.2.1915

Went to camp. Col. (Braund) and Bob to dinner č (with) Col. Nash. Escorted us home.
Mjr (Major) Gordon gave us some chocs. (chocolates) for Wyse.

The Pyramid's Cinema. AWM, PUBS002/003/002/001/007

171. The Mousky was a famous Cairo bazaar and is now known as El Mosky. Bazaar
172. Sault's Restaurant in Cairo
173. Seven Cairo cinemas are listed by the Australian War Memorial. Cinema Violet, Grand Cinema Olympia, Obelisk Cinema, Empire Cinema, Salle Kleber, American Cosmograph, Australian Imperial Force, Mena Camp, The Pyramids Cinema
174. Saqqara, an Egyptian village 24 kilometres southwest of Cairo, contains numerous pyramids

18.2.1915

On duty. Bob called in evening.

19.2.1915

Bob and Col. (Braund) called. Went to Cairo č (with) Bob and Col. visited Mousky (bazaar) and old house.

20.2.1915

Duty. Smallpox in camp.

21.2.1915

Camp p.m. Bob Kerr[175] arrived and had dinner č (with) Bob in his tent. Col. (Braund) took Fill and I to Sphinx in moonlight. Arrived home 11 p.m.

22.2.1915

On duty.

23.2.1915

Visited old Cairo Coptic[176] and Jewish[177] Church etc. Had dinner in Cont. č (with) Bob and Col. (Braund) also mosque of Aruru (sic?). 640 BC.[178]

24.2.1915

On duty.

25.2.1915

On duty a.m. Visited zoo.[179]

26.2.1915

Rubber (sic?) Sp. On duty.

27.2.1915

Off duty p.m.

28.2.1915

On duty. Tents coming down.

---

175. Louisa's cousin, Robert Alexander Kerr, born 1895
176. Possibly Saints Sergius and Bacchus Church, which dates back to the 4th Century
177. Possibly Ben Ezra Synagogue, the oldest Jewish temple in Egypt
178. The date may refer to the conquest of Egypt by Muslim Arabs
179. Giza Zoo, opened in 1891

March 1915

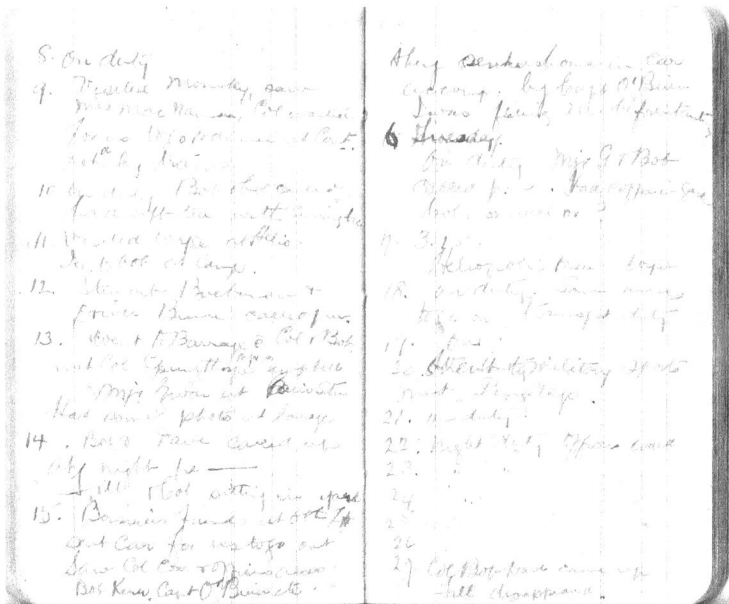

Entries for 8-27 March 1915

1.3.1915 March
Bob and Col. (Braund) took us out.

2.3.1915
On duty.

3.3.1915
Out p.m.

4.3.1915
On duty.

5.3.1915
Went into Cairo. Wyse to Heliopolis.[180]

6.3.1915
On duty.

---

180. Sister Wyse was stationed at No. 1 AGH, at Heliopolis

Heliopolis Palace Hotel, location of No. 1 AGH. AWM, H02266

7.3.1915
Church Parade at camp. Met various people. Off duty. Had morning tea at camp. Col. Nash called.

8.3.1915
On duty.

9.3.1915
Visited Mousky (bazaar). Saw Miss MacNamara. Col. (Braund) waited for us to go to dinner at Cont. (Continental). Ret'nd (returned) by train.

10.3.1915
On duty. Bob and Col. (Braund) called. Had aft. (afternoon) tea with sewing bee.

11.3.1915
Visited Wyse at Helio. (Heliopolis)
Illegible to Bob at camp.

12.3.1915
Stewart Buchanan (sic?) and driver Bruce (sic?) called p.m.

13.3.1915
Went to Barrage with Col. (Braund) and Bob. Met Col. Springthorpe, Mjr. (Major) Campbell.
Mjr. Gordon at Cairo Stn. Had some photos at Barrage.[181]

---

181. The Delta Barrage Dam was finished in 1862 to assist navigation and irrigation on the Nile

Le Caire — Entrée du Mousky

No. 672 Veguios & Zachos. Cairo & Luxor

Entrance to the Mousky Bazaar

14.3.1915

Bob and Dave[182] called up at night *illegible -*
Fill and Col. (Braund) sitting in yard.

15.3.1915

Boissier's friends at 6th L.H. (Light Horse?) sent car for us to go out. Saw Col. Cox and officers, also Bob Kerr, Capt. O'Brien etc. They sent us home in car accomp. (accompanied) by Capt. O'Brien. I was feeling ill before starting.

16.3.1915 Tuesday

On duty. Maj. G (Major Gordon) and Bob called p.m. Had coffee in garden. Col. (Braund) not well or ?[183]

17.3.15

Heliopolis to see Wyse.

18.3.1915

On duty. Some nurses to go on transport duty.

19.3.1915

On d. (On duty?)

20.3.1915

Went to military sports. Met Pengelleys.[184]

182. Possibly David McNeil Heugh, a cousin of Robert and Louisa's
183. This is Louisa's own use of the question mark
184. Charles Rosenthal, who joined the AIF in 1914 and was promoted to Brigadier General in 1916, noted that Mrs Pengelley and the two Miss Pengelleys were at Mena Camp and that their uncle, who lived at Maadi, was active in his help for both officers and men.

21.3.1915

On duty.

22.3.1915

Night duty. Officers ward.

23-26.3.1915

-

27.3.1915

Col. (Braund), Bob, Dave came up. Fill disappeared.

28.3.15 Sunday

Went river picnic č (with) 2nd Battaln. (Battalion) to Bedrashon.[185] Donkey ride. Memphis[186] etc. Very late.

29.3.1915

On duty. Bob called on way to the camp č (with) American ladies.

30.3.1915

Bob admitted to officers ward č (with) pneumonia.[187] We went to Mousky in afternoon.

31.3.1915

On duty.

The Delta Barrage Bridge across the Nile at Cairo. AWM, JO2249

185. Badrashin (30 km southwest of Cairo) is an area that contains the village of Saqqara
186. Memphis was the site of Ancient Egypt's capital, located at Badrashin
187. The Australians at Mena Camp were under canvas and the hot, dry and dusty desert climate
- hot by day, cold by night – led to respiratory issues in many instances

April 1915

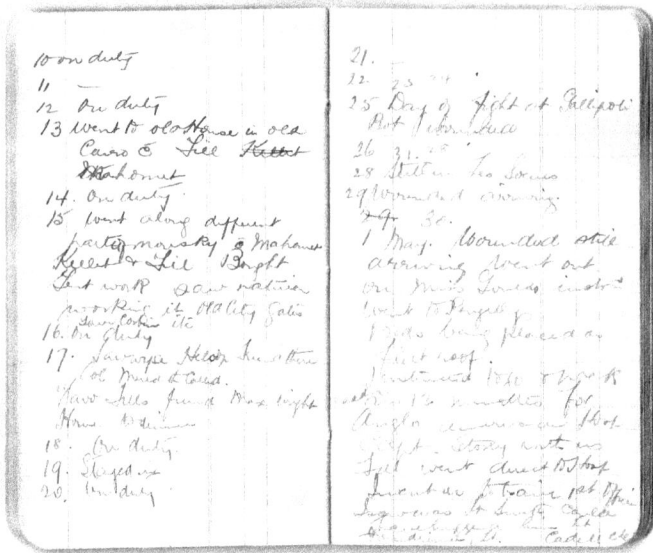

Entries for 10-30 April and 1 May 1915

1.4.1915 April 1st

Supper at Sphinx. Cairo for cakes. Colonel Springy (Springthorpe). Late home and in starting. Late coffee, supper, everything. Miss Gould wrath and let us know.

2.4.1915

On duty. Fill, Col. (Braund) and Lt. (Lieutenant) Pain (sic?) went up pyramid, too tired myself.

3.4.15

Night duty. 2nd Batt (Battalion) leaving. Bob still in hosp.

Went to camp for dinner. Cook[188] Wake and Fill and I had to leave early and walk home alone in dark, was very much annoyed.

Went with Cook and Sinclair Wood[189] to see Batt. (Battalion) march out. Motored ahead. Scattered batt. (Battalion) on road. Last we saw of them.

4.4.1915

On duty.

---

188. Elsie Shepherd Cook was married to George Sydney Cook, an officer in the 2nd Battalion and the son of Joseph Cook, Prime Minister of Australia from 24 June 1913 to 17 September 1914
189. Janet Sinclair Wood was a South Australian nurse who also travelled on the *Kyarra*

**Ethel Julia (Nellie) Gould
with Captain Horace Kingsmill.**
Gould was Matron of No. 2 AGH.
AWM, P07973.006

5.4.1915
Dusty day. Khamseen.[190] Artillery leaving. Capt. McGee very sad

6.4.1915
On duty

7-8.4.1915
-

9.4.1915
Drove to Khedival tombs.[191]

10.4.1915
On duty.

11.4.1915
-

12.4.1915
On duty.

13.4.1915
Went to old house in old Cairo.[192] With Fill. Mahomet.[193]

14.4.1915
On duty.

15.4.1915

Went along different part of Mousky č (with) Mahomet, Kellett and Fill. Bought tent work, [194] saw natives working it. Old City Gates Tour *illegible* etc.

16.4.1915

On duty.

17.4.1915

Saw Wyse Heliop. (Heliopolis). Friend there. Colonel Meredith[195] called.

190. Khamsin is a dry hot wind affecting Egypt
191. Louisa is referring to the tomb of the Khedives in Cairo. The Khedives had ruled Egypt in the late 19th Century
192. The historic area of Cairo
193. Possibly a guide or assistant
194. Khayamiya or tent work is an appliqued textile, depicting Egyptian scenes and figures
195. John Baldwin Hoysted Meredith was a doctor and soldier. He had likely served with Robert Scobie in the Citizen's Bushman's Contingent in the Boer War

Saw Fill's friend Max Wright (sic?) home to dinner.

18.4.1915

On duty.

19.4.1915

Stayed in.

20.4.1915

On duty.

21-24.4.1915

-

25.4.1915

Day of fight at Gallipoli. Bob wounded.

26, 27.4.1915

(no entries)

28.4.1915

Still in Les Soeurs (sic?)

29.4.1915

Wounded arriving.

30.4.1915

(no entry)

Some of the first wounded men from Gallipoli, at Mena House Hospital.
The photograph was taken by Irene Read, wife of Dr William Read and sister of
Fred Stobo Phillips. Mrs Read visited Cairo for several months in 1915

May 1915

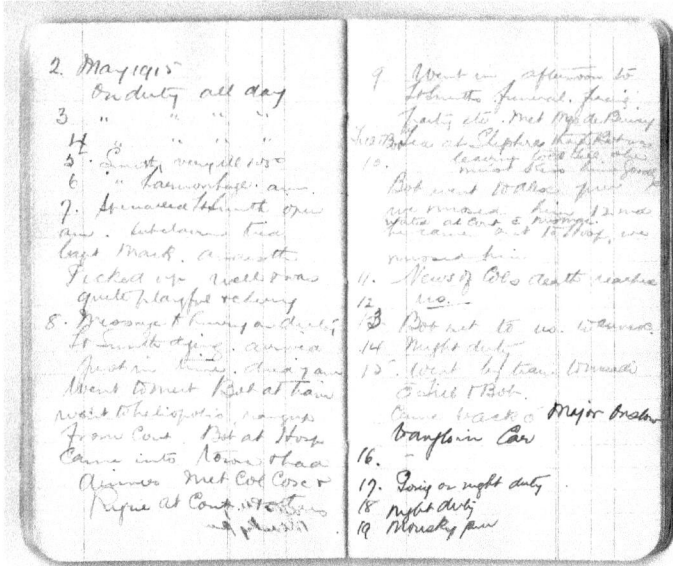

Entries for 2-19 May 1915

1.5.1915 May

Wounded still arriving, went out on Miss Gould's instruct. (instruction).
Went to Pengelleys.

Beds being placed on flat roof.

Instructed to go and pack ready in 12 minutes (sic?) for Anglo American
Hosp. Capt. Storey[196] with us. Fill went direct to Hosp. I went on to
train[197] 1st Officer I saw was Lt (Lieutenant) Smith,[198] could have hugged
him, Lt *illegible* Lt Cadell[199] etc.

2.5.1915 May

On duty all day.

3-4.5.1915

(marks may indicate as above).

196. John Colvin Storey was a surgeon with No. 2 AGH. Born in Ashfield in 1886, he had
travelled to Egypt on the *Kyarra*
197. Trains transported the wounded back to hospitals and nurses were deployed to provide what
treatment they could
198. Frederick George Smith, 2nd Battalion AIF, died on 8 May 1915, of wounds received at
Gallipoli on 27 April
199. Thomas Cadell was born in Goulburn in1894

5.5.1915

Smith very ill 105°

6.5.1915

Smith haemorrhage arm.

7.5.1915

Specialled[200] Lt. Smith open arm.
Subclavian tied.[201]

Capt. Mack anaesth (anaesthetic sic?).

Picked up well and was quite playful and cheery.

Frederick George Smith, Lieutenant,
2nd Battalion, died on 8th May 1915.
AWM, H13897

8.5.1915

Message to hurry on duty.

Lt. Smith dying. Arrived just in time. Died 7 a.m. Went to meet Bob at train. Went to Heliopolis, rang up from Cont. (Continental). Bob at Hosp. from cond. (condition). Came into town and had dinner. Met Colonel Cox[202] and Ryrie[203] at Cont. and others.

9.5.1915

Went in afternoon to Lt. Smith's funeral firing party[204] etc. Met Mqr (Marquis) de Busy[205] (sic?).

---

200. Specialling was a term used to indicate 1:1 patient care and keeping a patient in sight day and night
201. The subclavian arteries supply blood from the heart to the upper body
202. Charles Frederick Cox, commanded 6th Light Horse Regiment
203. Granville de Laune Ryrie was born in Michelago, NSW in 1865
204. The Australian War Memorial states that three rifle volleys were fired, in unison, three times as a mark of respect as a gesture of mourning at a soldier's funeral
205. Likely to be Sergius Mortimer Emmanuel Rouault De Longueville, 11th Marquis De Bucy, Captain, Headquarters Staff, AIF. De Bucy had served in the Boer War with the Johannesburg Mounted Rifles. Similarly Robert Scobie had served in the Boer War with the 3rd New South Wales Mounted Rifles

10.5.1915

Fill and Bob. Tea at Shepheards, thought Bob was leaving. Told Fill she must kiss him goodbye. Bob went to Alex. (Alexandria) p.m. We missed him 12 m.d. (midday). Waited at Cont. (Continental) č (with) Mrs Mac. He came out to Hosp. We missed him.

11.5.1915

News of Col's[206] (Braund) death reached us.

12.5.1915

-

13.5.1915

Bob ret. (returned) to us wounded.[207]

14.5.1915

Night duty.

15.5.1915

Went by train to Maadi with Fill and Bob. Came back with Major Onslow[208] to Anglo in car.

16.5.1915

-

17.5.1915

Going on night duty.

18.5.1915

Night duty.

19.5.1915

Mousky p.m.

20-22.5.1915

Night duty

---

206. Colonel George Braund was killed accidentally when failing to hear a sentry at brigade headquarters on 4 May 1915
207. According to the Virtual War Memorial, Robert Scobie was wounded early in the Gallipoli engagement when a bullet or piece of shrapnel injured his eyebrow and nose https://vwma.org.au/explore/people/791479
208. Possibly George Macarthur-Onslow, born in 1875, second-in-command of the 7th Light Horse Regiment

23.5.1915

Went to Ghezireh Gardens č (with) Fill, Bob and Bob Kerr, took photos, then into town to afternoon tea.

24.5.1915

(no entry)

25.5.1915

Tea at Groppi's … Mjr de Busy (sic?) and Bob.

26.5.1915

(no entry)

27.5.1915

Motored to Helouan.[209] Met Bob Kerr on road. Had afternoon tea there. After a scorching drive returned by motor somewhat late. Tyre broke going and coming.

28-29.5.1915

Had tea at Cont. (Continental Hotel) č (with) Bob.

30.5.1915

Out to dinner č (with) Bob.

---

209. Now spelled Helwan, c.30 km south of Cairo. The Al Hayat Hotel in Helwan was used as a convalescent hospital for wounded AIF personnel

June 1915

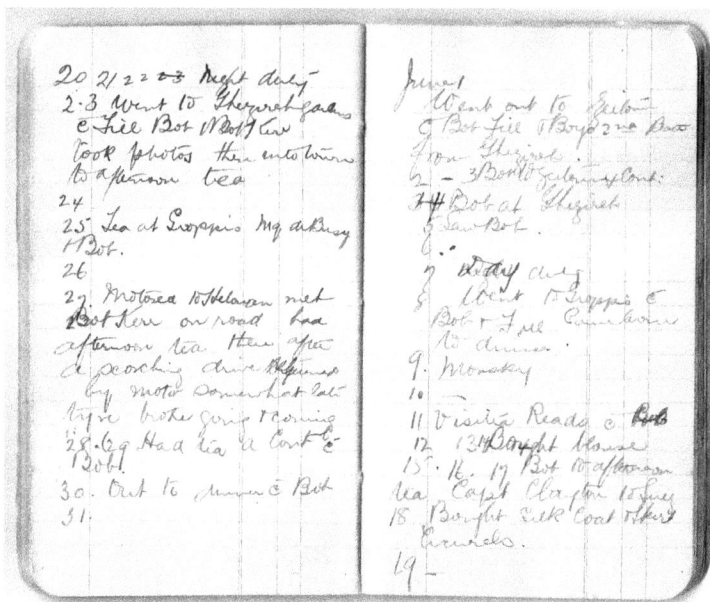

Entries for 20-31 May and 1-19 June 1915

1.6.1915 June
Went out to Ghezireh with Bob, Fill and boys 2nd Batt. (Battalion). From Ghezireh.

2-3.6.1915
Bob to Zeitoun and Cont. (Continental).

4.6.1915
Bob at Ghezireh.

5.6.1915
Saw Bob.

6.6.1915
(Marks may indicate as above).

7.6.1915
Day duty.

8.6.1915
Went to Groppis č (with) Bob and Fill. Came home to dinner.

9.6.1915

Mousky.

10.6.1915

-

11.6.1915

Visiting Reads[210] č (with) Bob.

12.6.1915

Bought blouse.

15.6.15

Bob to afternoon tea. Capt. Clayton[211] to Suez.

18.6.1915

Bought silk coat and skirt Cicurels.

19.6.1915

-

20.6.1915 Sunday. June.

Rabeah (sic?) Drive.

Saw wonderful motor wedding turn out. Dinner Dr. James.

21.6.1915

-

22.6.1915

Morning tea Groppi's č (with) Bob and Bob Kerr.

23.6.1915

-

24-25.6.1915

(no entry)

26.6.1915

Zoo č (with) Fill and Bob. All very weary. 27.[212]

28.6.1915

Bob to Alex. (Alexandria) a.m. Met Major Farr[213] at lunch Groppis. Drove in motor car to Zeitoun head office č (with) Bob. Home. Saw Bob off on train.

---

210. This is likely to be Dr William Read and his wife, Irene. She was related to Louisa by marriage
211. Possibly Hector Joseph Clayton, born 1885, wounded at Gallipoli in May 1915 and served as an embarkation officer in Egypt until mid-1916
212. The date is attached to the entry for 26 June
213. Walter Percy Farr, 1st Light Horse Regiment was born 1889

29.6.1915

To return Ghezireh but did night duty for 1 hour, out on river č (with) Sst (Sisters) Saunders, Froler (sic?) Llewellyn etc.

30.6.1915

Went to Ghezireh a.m. no sleep, glad to be back. Off duty p.m. Went out, went out (repeated) in afternoon to Cairo. Came home in motor unit č (with) Fred Phillips.

Café Groppi was located on Soliman Pasha Square
(now known as Talaat Harb Square) in Cairo. Newsweek states that it was founded in 1909 by Giacomo Groppi, a Swiss pastry chef.[214] It was an alternative to the more traditional male-only coffee shops.

214. Salama, V. (2012). 'The Enduring Charm of Café Groppi in Cairo.' Newsweek, 20 August 2012

July 1915

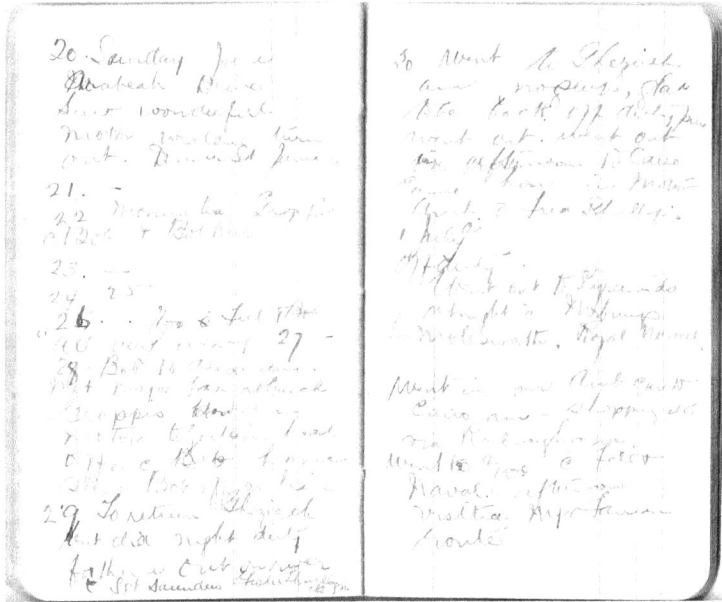

Entries for 20-30 June and 1 July 1915

1.7.1915 July

Off duty.

Went out to Pyramids at night č (with) Norburys (sic?) Lt. (Lieutenant) Molesworth,[215] Royal Naval.

Went in on amb. (ambulance) car to Cairo a.m. – shopping etc. via Railway House. Went to zoo with Fill and Naval[216] afternoon. Visited Mjr (Major) Farr *en route*.

2.7.1915 July

Out č (with) Naval to Heliop. (Heliopolis) to look up Don[217] and Wal.[218] Naval arranged to call for me p.m. Brought Molly[219]. Fill did not tell me. Was much annoyed. Saw them and went to Groppis, home and out on river. –

---

215. Lieutenant Molesworth, Manchester Regiment. The poet Wilfred Owen was a member of the same Regiment
216. Lieutenant Harold Sawyer, Hawke Battalion, born 1892
217. Donald Douglas Scobie, born 1893, Louisa's nephew, was in the Signallers Unit of the 18th Battalion, 5th Brigade
218. Wallace Wilfroy Scobie, Donald's elder brother, 18th Infantry Battalion, born 1891
219. This appears to be a nickname for Lieutenant Molesworth

10.15 p.m. Comdr. (Commander) Dalgliesh (sic?) and Sister Bannister in another boat. Lovely time tacking on river.

3.7.1915

On duty 11. Naval rang up a.m. to say goodbye. Read letter from him.

4.7.1915

Rested p.m.

5.7.1915

No entry

6.7.1915

Went to zoo č (with) Fill. Met Lt. *illegible*. Then out to Heliop. (Heliopolis). Morn. tea obelisk.

7.7.1915

No entry.

8.7.1915

Not out.

9.7.1915

Wounded arrived.

10.7.1915

Went for motor drive with Lts (Lieutenants) Molly and Boulster,[220] Gurka to Matarieh[221] saw church[222] and obelisk.

11.7.1915

-

12.7.1915

Went into town p.m. Half expected Capt. Campbell[223] to call. Did not arrive. Molly and Gurka called at dentist.

13.7.1915

Went to Pyramids č (with) Capts. Stevens and Campbell. *Illegible* a dead head. Rode from Sphinx on camel. Light ref. (refreshment) at Greek restaurant. Home to duty 5 p.m.

220. Lieutenant Boulster, Gurkha Regiment
221. Al Maṭạrīyah, c. 10 km northeast of Cairo
222. The Holy Family Church in Maṭạrīyah, contains murals which represent the journey of the family from the Massacre of the Innocents to Upper Egypt
223. Captain Campbell of the Ayrshire Regiment had been a patient of Louisa's at the Anglo American Hospital

Louisa, second from right, with Wilhelmina Fillans, left and unidentified sister.
The soldier at left may be Captain Stevens or Captain Campbell.[224]

14.7.1915

Off duty to pack for Alex. (Alexandria) and train duty.[225] Said goodbye to Pts (patients) in No. 11. Byron Ross in both waved his hand.

Lft (Lieutenant) Davidson[226] in charge. Travelled č (with) Capt. Le Mesurier, [227]very nice man. Went (sic?) Alex (Alexandria) by motor. Taken to Caracol [228] main guard (sic?) and met Capt. O'Farrell, who gave us address Miss Smythe Quarantine Depot, The Arsenal. Saw old Col. *illegible*. Kicked out etc. Arrived here. Met by Miss Smythe who is exceedingly nice.

Large airy home. (2 four poster beds) overlooking harbour, mosque, native quarters. Went to town p.m. Met our predecessors, Wallace and Backhouse who stayed for night.

224. Collection of Dinah Toohey
225. Hospital trains were deployed with medical staff, who could provide immediate intervention as the wounded were transported to hospital centres
226. Possibly William Davidson, of the 1st Australian Infantry Battalion, who died of wounds at Gallipoli on 19 August 1915
227. Haviland le Mesurier was born 1856, and served with 8th Australian Commonwealth Light Horse
228. Caracol was the British Garrison Headquarters in Alexandria

15.7.1915

No duty. Wallace and Backhouse left for duty No.19 Hosp.[229] Went over to trains to inspect, very nice motor launch trip over. Went to town p.m. Saw Hosp. ship *illegible* 12.

16.7.1915 Friday

Hosp. ship[230] in. Miss Smythe and I went over in launch č (with) Lady (intentional blank space) whom I escorted aboard the *Asturias* H.M.H.S.[231] Many very bad cases aboard. Wards looked very comfortable. Apparently territorials[232] aboard. Came home to lunch. News arrived that Nos. 1 and 3, Fill, trains to go. No. 2 not going. Stayed in, read, rested, sewed, wrote diary, talked to Miss Smythe etc. Fill (Fillans) still away on train at Cairo o'night (overnight).

This is the last entry in the diary until a single page entry in September 1919. Over the remaining years of the war, mention of Louisa's friend, Wilhelmina Fillans ends in 1916 but their attestation papers show they were both frequently at the same postings so it is likely their friendship and mutual support continued.

229. No. 19 British General Hospital, Alexandria
230. Hospital ships covered one stage of the wounded's transfer from Gallipoli to a base hospital
231. His Majesty's Hospital Ship (HMHS) *Asturias* was one of the largest cross-channel hospital ships. It was attacked and wrecked by a German submarine on 21 March 1917
232. The Territorial Forces Nursing Service (TFNS), established in 1908, was deployed as part of the British Army Medical Service

Louisa's war in letters and reports home

**HELP THE
AUSTRALIAN
COMFORTS FUND**

Poster for the Australian Comforts Fund, AWM ARTV06422

Louisa's diary ends abruptly at this point, not to be taken up again until September 1919. But her letters home[233] and those of others who mention her or Robert Scobie provide some record of her reflections and experiences over the next 3-4 years. These and details from her attestation 'papers' help to establish a chronology of her movements and experiences in 1915 and in the following years.

In addition, reports of some of Louisa's letters, often to her sisters, were published in regional newspapers, like the *Maitland Daily Mercury*, (referred to as the *Mercury* in the following pages). Some letters occasionally found their way into other regional and city journals. The first of these relates to incidents mentioned in the last months of her diary.

One of the first mentions of her came from Fred Phillips, Louisa's nephew by marriage, who was serving with the AMC at Mena. Writing to his sister on the last day of January 1915,[234] he spoke of the discomforts of doing night duty in the hospital tents at Mena.

"Sister Stobo has been very kind to me," he wrote. "She is a very fine woman."

---

233. A small selection of Louisa's letters are held by Robert Scobie's great-granddaughter, Dinah Toohey, while a further letter is held by Mary Thom, granddaughter of Elizabeth Roberts' (née Scobie)
234. Letter to his sister, dated 31st January 1915, Mitchell Library, MLMSS 2836 (K22194/Folder 8)

In June, the *Daily Telegraph*[235] reported that in the previous month, Louisa had been nursing at the Anglo-American Hospital, Ghezireh Palace. Sister Agnes Shearer, who had also embarked on the *Kyarra* and was attached to No. 2 AGH described the hospital in some detail.[236]

> "The hospital was in a beautiful position on the banks of the Nile, and it had very large airy rooms, well-equipped, and the food was very plentiful ... There were beautiful gardens around the hospital where the patients were able to walk."

No. 2 Australian General Hospital moved to the Ghezireh Palace (pictured)
from June 1915 to treat casualties from the Gallipoli campaign. The original
Mena House Hospital handled the overflow. AWM A01624

In Louisa's letter, she wrote of the number of patients then under treatment – 500 – and their good spirits.

> "You would be astounded she continued, "to know how bright and jolly all the men are, always joking, with bright, shiny eyes. Really it is marvellous to think they have been in the thick of it, and yet almost all are wishing they were fit to go on again and give the Turks one back. The dash and initiative of the Australians come in for commendation. The Australians have made a name for themselves."

235. 'Always Jolly', Daily Telegraph, 12 June 1915, p.10
236. Interview with Sister Agnes Shearer, AWM 41: 1072. Interviews containing accounts of Nursing experiences in the AANS [Australian Army Nursing Service], p. 107

More details come from other sources, such as a 20th May letter[237] to Mrs. J. E. Chant of West Maitland, from her son, Sergeant Gordon Chant.[238] He gives details of his injuries and the landing at Gallipoli and the Maitland friends who have died. In hospital, with a Maitland friend in the next bed, he wrote:

> "Major Scobie and his sister (Mrs. Stobo) have been in to see us a couple of times. By Jove, they are both fine people in every way. They brought in a splendid assortment of fruit and other articles the other night and another day. A lady friend met the Major quite by accident. She didn't know he was wounded. Anyway, next thing happened was she brought in to the second battalion a great pile of jelly, tinned fruit, oranges, and tomatoes. Most of the Cairo people are very good to the wounded. Major Scobie's wound is doing well, but he is still weak from loss of blood. He will be returning to Gallipoli at an early date now."

By late June,[239] the *Mercury* was reporting a May letter from Louisa in depth. She wrote of the pressure on medical and nursing staff at the arrival of large numbers of wounded soldiers from the fighting at Gallipoli.

> "I had a letter from Sister White,"[240] wrote Louisa, "in which she said they had only been in Alexandria two hours when the wounded began to arrive in hundreds. The building made use of for a hospital had been a school and was most inconvenient. They then only had six sisters to one thousand patients, There are still between 2000 and 3000 at Mena House, 1000 to 5000 at Ghezireh Palace,[241] 700 at Luna Park,[242] and from 8000 to 9000 at Heliopolis Palace. A number of the patients have minor wounds, the more serious are being kept in Alexandria."

237. 'Australian Heroes, The Maitland Boys – Wounded Soldier's Tribute', Maitland Daily Mercury, 24 June 1915, p. 3
238. Gordon Leslie Chant, 2nd Infantry Battalion. He was killed in action in Belgium on 15 September 1916
239. 'Hospital Work in Egypt. A Nurse's experiences,' Maitland Daily Mercury, 25 June 1915, p.3
240. Possibly Jessie McHardy White, who served at No.1 Australian General Hospital in Egypt
241. No.2 Australian General Hospital was located in the Ghezireh Palace from June 1915
242. The buildings and grounds of the Cairo Luna Park had been adapted as an auxiliary hospital in mid-January 1915

The letter carried details of Robert Scobie, who was in the convalescent annexe of the hospital on May 23rd but expecting to return to the front soon. Louisa wrote of the need for socks and other 'comforts' for men at the front and of the wounding or death of 29 officers from the 2nd Battalion.

> "An English officer told me," she continued, "there were wire entanglements beneath the water when they were landed. One ship had lighters lashed alongside with the idea that when the ship was run ashore the lighters would swing ahead and form a landing place. This scheme failed: the men meanwhile jumped on to the lighters (pontoons or some such) and were shot in large numbers by machine gun fire."

The letter also carried news of the death of her friend, Major Gordon, a frequent companion in the weeks preceding the landing. It seems, she wrote, that he was leading his men,

> "...in his usual jolly style, saying, 'Come, along boys, who's for' six-penny seats,' when he gave the sudden order, 'Down! down!' but was standing himself when he was shot; and as the boy who related this story remarked, 'died like a lion.' Colonel Braund, it is related was leading the boys to the trenches when a sniper shot him through the heart."[243]

A report in the *Mercury*[244] from Private William Dilley[245] tells of the death of other Maitland soldiers, including Second Lieutenant Eric Sölling.[246] Sister Sölling was by now nursing in Egypt and it fell to Louisa to tell her of her brother's death. Dilley was being nursed by Sister Sölling and was a witness to the event.

Some insight into Louisa's character can be gained from a report relating to her experiences in *The Tamworth Daily Observer*.[247] In it she speaks of the adventures of Lieutenant Harold Sawyer, of the Royal Naval Division,[248] who

243. Initially, there was misinformation about Colonel Braund's death
244. 'Wounded in Action', Maitland Daily Mercury, 10 June 1915, p. 4
245. William Dilley, died 3 May 1917 in France
246. Eric Martin Sölling was born in Hexham 1893
247. 'A Naval Man's Experiences, Quartermaster blown out of dug-out,' Tamworth Daily Observer, 28 August 1915, p. 4
248. Harold Algernon Sawyer, Royal Naval Division, Hawke Battalion, born 1892. Sawyer was later (23.7.1918) awarded the Military Cross for conspicuous gallantry and devotion to duty. https://www.naval-history.net/WW1NavyBritishLGDecorationszzMC.htm

was captured in Holland but managed to escape and make his way back to England. She was nursing English, Scottish and Australian soldiers at the time, one of whom, though not wounded, was suffering from shellshock. Lieutenant Sawyer was a particular favourite, whom Louisa speaks of affectionately in a letter of 7 June.[249] Sawyer had recently taken Louisa and her friend, Wilhelmina Fillans, to the zoo and later that evening, sailing on the Nile in a felucca.[250]

> "We had a most picturesque old boatman," she wrote. "He had a white native costume on, well blue-bagged and baggy about the pants. Every time he dived from one end of the boat to the other, Lt. used to call out 'Antonio! Mind your trousis.' We all laughed, and the old boatman would chuckle away too."

Sawyer had written to his mother that, as nurses, Louisa and 'Fill' were of "the good old sort". Louisa was impressed by the 23-year-old who was articled to a lawyer in peacetime, had studied political economy and could speak French, German and Spanish.

> "I simply adore the kid!" she wrote.

Kirsty Harris has described the central role nurses came to play in soldier's lives.[251] She cites historian Janice Gill, on the psychological value of nurses and how the 'comfort' aspect of nursing grew exponentially in these conditions. Harris also quotes historian Katie Holmes on the ability of nurses to adapt their manner of relating to patients based on the patient's age, health and the nature of their wounds. She further notes that, for patients, female nurses created a psychological link to home. It seems reasonable to suppose that Louisa's focus on Maitlanders among the wounded would have enhanced this link, both for her patients and their families.

Louisa's August 1915 letter also speaks of the conflicting stories surrounding Colonel Braund's death and includes comments on the Australian soldiers, sent home after disgracing themselves and the AIF in the infamous Battle of the Wazzir.[252] She showed compassion for these men and believed they should

---

249. Letter dated 7 June 1915, collection of Dinah Toohey
250. A felucca was a wooden boat, with a single sail
251. Harris, K. (2013). 'All for the Boys,' The Nurse-Patient Relationship of Australian Army Nurses in the First World War, in First World War Nursing: New Perspectives, edited by Alison S. Fell and Christine E. Hallett, Routledge Studies in Modern History Series, p. 71-87
252. The Battle of the Wazzir was a cumulative term for several forays by Australian soldiers into the Red Light District of Cairo. The worst took place on 2 April 1915 and over 300 soldiers were sent back to Australia in disgrace

have been given a second chance as their lives would now be lived "under a cloud". She reports on the black humour of the wounded and says they were "as proud as Punch of their wounds and as bright as possible." A letter dated 21st June[253] records Louisa's contact with some Maitland soldiers and speaks of the frustration many felt at their inactivity.

> "A fresh lot of wounded arrived last week," she wrote, "mainly Britishers. They say that the Australians are becoming tired of inaction, though I guess they get plenty. We met a trooper from the 6th Light Horse lately. He says one would laugh at the cuts some of the men present with almost no pants left, they have made them so short. They cannot get a wash except by going into the sea should a lull occur as water is very scarce where they are (though in some parts of the Peninsula it is to be had for the digging)."

In a July letter to her sister May,[254] Louisa mentions that Robert had lost a lot of blood while waiting, wounded, in the firing line to be rescued. A report in the *Mercury*[255] gave further details of the wound. He had a small piece of strapping across his nose, but otherwise, looked well. It was evident to Louisa, however, that her brother had been through considerable strain.

> "You should hear the boys speak of his bravery," she wrote. "They say he was not even fit to go ashore on the day he landed, and yet he went, feeling weak, and helped the boys, and led them on. One boy said, 'If ever there was anyone worthy of the name of man it is Major Scobie.' They say Bob led 20 men onto a ridge, through and into what the boys said was simply, 'a hell' of shrapnel. Almost everyone returned, and they accounted for a lot of the Turks. Bob says the boys were marvellous, never refused to go ahead when asked, and, in fact, did not require asking, as a rule, simply forged ahead."

253. 'Troops at Gallipoli, What they require, Red Cross Arrangements', Maitland Daily Mercury, 28 July 1915, p. 3
254. Letter to May Scobie, 14 July 1915, collection of Dinah Toohey
255. 'The Wounded in Egypt, a Nurse's Experiences', Maitland Daily Mercury, 9 June 1915, p. 6

Robert Scobie, Lieutenant-Colonel, (front row, 3rd from left)
commanded the 2nd Battalion after George Braund death. AWM, P03796.002/

The letter carried news of Maitland local, Lieutenant David Heugh;[256]

> "one of the first of the East Maitland High School 'old boys' to lay down
> his life in the Empire's supreme struggle. Poor dear old Dave was shot
> through the liver and had a bad time. Thank goodness, though, he had
> mates aboard. Among them were Father (Captain) McAuliffe,[257] who
> was very fond of him, and used to ease him by supporting him in his
> arms at times, also Lieutenant Cadell,[258] of Newcastle, who was slightly
> injured, and is writing to Dave's father. Dave was a great favourite with
> anyone he knew."

She mentions that among men she nursed were Lord Loughborough of the
Coldstream Guards "a fine big boy, with no nonsense" and Sub-Commander
Price, a nephew of Mr. Winston Churchill, "a fine, pleasantly spoken, intelligent
man, of about 35 years of age."

256. David McNeil Heugh, born Ballina NSW 1891, died 29th April 1915 on board the hospital
ship, Derflinger. Heugh was Louisa's second cousin
257. Father (Major) Edmund McAuliffe was born in Ireland in 1881
258. Lieutenant Thomas Cadell aged 20, died of wounds at Gallipoli on 22nd June 1915

"Commander Price," she wrote, "says he does not think that the 'Tommies'[259] would have done as well, for the Australians did not want to be led, but simply went ahead, and each did his share. Captain Dr. Butler,[260] of Brisbane, led a charge; some say he will get the V.C.; a priest led another, also one of the company's cooks."

Two letters from July 1915 record some of her last contact with Bob before he was killed in action some time between 6th to 8th August.[261] In one she writes that he has sent her a wire, saying he was leaving for Gallipoli that day (7 July).

"I was hoping that he might be kept at Alexandria," she wrote, "though he is looking very much better than when he arrived at Cairo. I went to Cairo station, to see him off the day he left for Alexandria, and we can only trust he may return again."

In another letter, dated 14th July,[262] she writes,

"There is no news from Bob yet; but somehow I feel he will be all right. He looked ever so much better. When leaving he said, 'If the boys were here, they would be saying 'Daddy's going to the war to kill Germans'. Bob is quite pathetic over the youngsters. He simply dotes on them. After he was wounded, he lost a lot of blood, and said all he wanted to do was to sit down somewhere, but he thought of Flora (his wife), and struggled on again."

At this time, Louisa was working on the hospital trains in Alexandria and one letter[263] records her impressions of wounded Turkish prisoners of war.

"Three trains went up to Cairo today," she wrote, "one lot being mainly wounded Turkish prisoners. Poor fellows, some of them were badly injured, and they wanted their wounds dressed too, for the ship they came on was not a hospital ship. Many of our men have had to be brought over that way too. I helped to get the men seated in the trains, but did not go up. Miss Smyth took my place for she could speak the Turkish language. The remainder of the

259. 'Tommies' was the nickname for British Army soldiers
260. Arthur Graham Butler was born in Kilcoy, Queensland, 1872. He was the only medical officer to receive the Distinguished Service Order at Gallipoli
261. 'Local News', Maitland Daily Mercury, 28 August 1915, p. 4
262. Letter to May Scobie, dated 14 July 1915. Collection of Dinah Toohey
263. 'News from Egypt', Maitland Daily Mercury, 27 August 1915, p. 4

patients were Indians, English, Scotch and Australians. One cannot realise that so many really fine fellows have gone, and often incorrect details are received. One hears about three stories about the one person. Shiploads of reinforcements are leaving here of late. It is to be hoped this warfare will soon cease for it seems to be such a dreadful waste of lives. Our men do not blame the Turks. They say they attend to our wounded well."

A wounded Turkish prisoner receiving attention at the 1st Australian Field Ambulance Dressing Station. AWM, A03770

It was the Sydney *Sun* [264] that brought further news of Louisa on 30 July, reporting a letter she had sent. In it she protested at the current arrangements for the distribution of Red Cross 'comforts' sent to the troops.

> "A responsible officer should be placed at each hospital," she wrote, "with two or three assistants, to see that each man receives all necessary articles of clothing, toilet necessaries, and a little pocket money, for, as I think I told you, they come in with absolutely nothing as a rule but their bloodstained and torn garments. And, for goodness sake, keep up the collections, for the war is not yet ended, and our men do and will require all the help you can give them. Many of the Red Cross people, also private

264. 'Nurse's Appeal, Keep Giving', The Sydney Sun, 30 July 1915, p. 5

citizens, give and give to the men, but even so the boys want a little cash to get little extras. The paymaster has been instructed not to pay convalescent soldiers in hospital, so what are they to do, unless their own people can do something in the matter? Anyhow, the men are as brave as lions, and as bright and cheery as possible, and seldom, if ever, complain about anything."

Red Cross boxes could contain a pipe, tobacco, cigarettes, chocolate, playing cards, matchbox and a handkerchief.
AWM, REL32973

The *Mercury* reported that food was plentiful, but that, in the main, the men required socks and changes of under linen.

"An officer who was there says thick cakes of chocolate are very good to send to the Peninsular. He bought £30 worth for his men."[265]

After Lieutenant-Colonel Braund's death, Robert Scobie became the commanding officer of the 2nd Battalion. Young James Larkin, a newly enlisted Private with barely any training or experience of war, left vivid details of Scobie's death.[266] Both had taken part in the Battle of Lone Pine from 6th to 10th August 1915, as part of the A.I.F. August offensive. This was a diversionary attack on a site south-east of Anzac Cove, with the objective of taking and holding the Turkish line. Over 2000 Australian troops died or were injured, with more than 6000 Turkish troops wounded or losing their lives.[267]

265. 'Troops at Gallipoli. What they require'. Red Cross Arrangements. Maitland Daily Mercury. 28 July 1915, p. 3
266. James (Jimmy) Larkins, Virtual War Memorial, Australia, https://vwma.org.au/explore/people/351265
267. Battle of Lone Pine, 6-10 August 1915, Australian War Memorial, https://anzacportal.dva.gov.au/wars-and-missions/ww1/where-australians-served/gallipoli/august-offensive/lone-pine

Robert Scobie with his wife Flora, sons, Kenneth, (at left), Roden and daughter, Jessie Agnes[268]

Larkin was badly wounded trying to capture one of the trenches. He lay injured and unprotected until the morning of 8th August, begging to be put into the trench away from heavy gunfire. He credits Robert Scobie as saving his life when he did so with the help of another soldier.

> "Captain[269] Scobie and his mate were the only two brave men there," he wrote. "They came out in the thick of the firing to save their comrades … The last time I saw Captain Scobie alive was when he left me in the trench. I asked him how many trenches we had captured and he said, 'we have captured three my boy.' … About a quarter past eight that night we were carried to the hospital ship. I asked one of the men carrying me where Captain Scobie was, and he said, 'you will soon see him now. He has been killed.' So, mother, the next I saw was Captain Scobie being laid by my side … Mother, you can thank the Almighty I am so far

268. Collection of Dinah Toohey
269. Larkin does not appear to know that Scobie was the commanding officer and so wrongly identifies him as Captain Scobie

safe, and I am satisfied that it was Captain Scobie who saved my life and many other poor wounded comrades as well."[270]

While there is some uncertainty about the exact time of Robert Scobie's death, official accounts describe his bravery and above all, his leadership and ability to inspire confidence in the men. Contemporary perspectives, however, suggest that Australian commanding officers lacked experience in trench warfare and that Scobie, along with the commanders of the 1st and 4th Brigade, engaged in the fighting "in a manner unseen in later operations."[271]

A later report in *Reveille*,[272] has left a detailed description of his last moments.

> "Scobie, with a wounded arm hanging in a sling, passed along the whole of the Second's position, and spoke to practically every man. ...From one post came an urgent call for more men to replace the heavy casualties. Scobie hurried along, as he always did, to the most threatened point, his intention being to see if this point was worth holding for the sacrifices being made, or to send in more men, if he decided in favour of holding. After viewing the position, he ordered a retirement of about 30 yards out of the communication trench that had been held so valiantly for such a time and was standing in the sap while the men withdrew, when he was killed. The withdrawal was delayed then until his body was removed. It was finally sent to the beach for burial close to his old comrade, Braund."

This report tallies with that left by Chaplin (Captain) William McKenzie,[273] who wrote to May Scobie after Robert's death.[274] A Salvation Army Chaplain, McKenzie was known as 'Fighting Mac' and was one of the first chaplains ashore at Gallipoli. He was awarded the Military Cross for distinguished services in the field.

270. James (Jimmy) Larkins Virtual War Memorial, Australia, https://vwma.org.au/explore/people/351265
271. Westerman, W. (2017), Soldiers and Gentlemen: Australian Battalion Commanders in the Great War 1914-1918, https://doi.org/10.26190/unsworks/2655
272. Herrod, Lt. Col. E. E., (1931), 'Trainers of Anzac: Braund and Scobie.' Reveille Vol.4 No.7; 31 Mar 1931 p29, 86-7)
273. McKernan, M., William McKenzie (1869-1947), Australian Dictionary of Biography, 1986
274. 'The Late Lieutenant Colonel Scobie', Newcastle Morning Herald and Miners' Advocate, 14 April 1916, p. 6

"I had difficulty in getting through and over the dead and wounded," he wrote, "and when I reached the first trench, where the men were falling by bombs all round me, killed and wounded, there stood your heroic brother, directing and encouraging his men in the hottest encounter of the bloody battle. Comparatively few of his officers and men were left, and they were falling rapidly, as the shells and bombs were bursting in fury on and in the trenches. This was the last view I had of the Colonel. He was a fine leader, and his men were particularly proud of him, and looked up to him with profound respect."

A further tribute by one of the men he commanded, was published in the *Mercury* late in August.[275] Private Sydney Leserve[276] spoke highly of all his commanders and made special mention of Lieutenant-Colonel Scobie. Leserve was with the 2nd Battalion when they landed at Gallipoli and recalled Scobie as being always in front of his men, leading them into bayonet charges.

"He was immensely popular with every member of the 2nd Battalion," Leserve wrote, "because he knew how to handle men and treated them in the best manner possible." Leserve also had contact with Louisa who he said was doing "splendid work with the wounded" at the Anglo-American Hospital at Ghezireh.

In the midst of this unrelenting suffering, Louisa continued to call for 'comforts', especially for 'Maitland boys' and she began to act as an intermediary for parcels sent from home that fell outside the ambit of official Red Cross parcels. For their part, the Australian Red Cross (ARC) sent c. 400,000 food parcels and c. 36,000 clothing parcels[277] over the duration of the war but these were sometimes held up in transit or delayed by lack of available transport from Australia.

Louisa's call for 'comforts' reflected the struggles those in Australia experienced regarding this issue. Historian Steve Marti, writing in History Australia,[278] outlines the difficulties patriotic organizations experienced in

275. 'Soldier's Tribute to deceased officers', Maitland Daily Mercury, 21 August 1915, p. 2

276. Sydney J. Leserve, No. 962, 2nd Battalion, appears in the lists of wounded for 15 May 1915

277. We will remember them, (2018) Australian Red Cross, https://www.redcross.org.au/stories/ihl/armistice-day

278. Marti, S. (2016). 'Knitted Together: Home, Front, and Voluntary Patriotism in Australia during the First World War', History Australia 13, No. 3. 386-381,

coordinating their efforts at State and Federal levels. Money and contributions raised locally were anticipated to find their way to local battalions but when the Australian Comforts Fund (ACF) was established in August 1916, local organizations were urged to send contributions to a union of state bodies, for sorting and distributing as they saw fit. Marti notes that such voluntary associations could be apprehensive about the destination of donations that were intended to support members of their own community.

The scale and type of 'comforts' which Louisa helped to manage is suggested in Table 2, which itemizes the goods collected by the Maitland Mayoress's Patriotic Committee, which were sent to Egypt for hospital use and for soldiers, fighting in the trenches.[279] "Goods were sent direct to Matron Gould and Sister Stobo, and through the War Chest Depot and Red Cross Society," wrote the Secretary of the Patriotic Committee.

Table 2: List of 'comforts' provided in 1915 by the Maitland Mayoress' Patriotic Committee

1091 pairs sox	450 mufflers
459 Balaclava caps	50 pairs gloves
447 suits pyjamas	628 flannel shirts
150 pairs sheets	665 handkerchiefs
479 glass towels	750 pillowcases
1784 assorted bandages	522 Turkish towels
250 milk covers	250 kit bags
28 mosquito sets	418 doz. safety pins
20 pkts. ab (absorbent) lint	20 pairs bed sox
12 doz. fans	1 dozen cushions
200 pairs underpants	164 rolls calico for bandages
420 eye bandages	567 face washers
8 utility kits	512 tins milk
164 rolls flannelette for bandages	364 cholera belts
268 packets cornflour	100 packets arrowroot
66 tins pt. (pot) barley	56 bottles Bovril
30 tins Nestle's food	2 gross matches
299 tins cocoa	1417 packets cigarettes
554 tins and cakes tobacco	360 pipe
10 boxes cigars, cigar-holders,	235 lb. boiled lollies
pipe cleaners, cigarette cases	20 boxes chocolate

280. 'Mayoress Patriotic Committee, Fine Record of Work', Maitland Daily Mercury, 24 August 1915, p.6

In September, she wrote to her older sister, Janet, a letter later published in the *Mercury*.[280] In it, she mentions how much in need the troops were of 'comforts' and diversions, and of a memorial service that was held for Robert Scobie.

> "Yesterday evening we gave high tea to about 100 Australians at the Quarantine Depot at Gabbari," she wrote.[281] "The boys enjoyed it immensely, and said it was like being at home. What is badly wanted at Cairo and Alexandria is a place—large roomy ... house preferably—arranged somewhat like a home, where the Australians can meet Australians. If only some nice, homely, entertaining people with hearts and thoughts for the welfare of the boys could undertake to sort of keep open house for the boys it would be a great boon to them and would do away with a lot of these rows and disturbance one hears of, for the boys have nowhere to go, and simply rove about full up with things in general, especially if they have been kept inactive for months on end.
>
> "One boy said to me last night that he would rather be shot than wait about as his company had been doing. He was simply downhearted over being kept here, though really the 4th L.H.F. Amb[282] is doing splendid work—hard too—unloading wounded and sick off the hospital boats. You would be pleased to see how gentle and thoughtful they are with the patients. Everyone speaks of them in the highest terms; they are mostly Queenslanders and South Australians."

Her suggestions show something of her skill for organisation and her practicality.

> "I think if it could be managed by entertainment committees to provide funds for entertainments," she wrote, "it would be a good thing for the boys. I suppose it would be too big a thing to rent a house as suggested, though I am sure there are many willing well-to-do Australians who would, and could do this, if only they knew what a boon it would be for the men. Even an Australian Club of

280. 'Australians in Egypt, A Nurse's Experiences,' Maitland Daily Mercury, 31 December 1915, p. 3
281. Gabbari Camp, Alexandria
282. The 4th Australian Light Horse Field Ambulance

this description, with a small weekly or monthly subscription I think, would prove to be satisfactory—a place where they could feel they were welcome; could be entertained; have baths, a cup of tea, etc., and where they could feel there were Australians to be met. Our boys do not care for the Tommies, and vice versa, I guess."

The letter also gives insight into the way nurses like Louisa used their own funds to buy 'comforts'.

"We had afternoon tea last Sunday at Gabbari for the boys," she continued. "They enjoyed it so much that we had yesterday's high tea, but as only a few of us give the tea we will be unable to continue at frequent intervals, for we will become stony broke. Even if we had a fund to provide cakes, tea, etc., and get local people to assist at these evenings we could manage to a small extent, though only a few local men would benefit through us."

Along with her commitment to providing some normalcy in the lives of the young men she nursed Louisa's letter also relates her own capacity to enjoy the uncommon sights around her in moments of reflection.

"Last week Miss Rundle and I went out to Montaga Castle, late Khedive's home", [283] she wrote. "It is built overlooking the sea and has an artificial harbour. The outlook is beautiful; we went upstairs at sunset; the soft and rich evening tints were glorious, reflected in the sea, with outlying bays, etc. Alexandria lighthouse in the distance; palm trees distinctly outlined against the beautiful sky. We came home by motor ambulance. Another day we went to Aboukir Bay—Nelson fame.[284]

"It is an uninteresting place as a place: nothing to do, but, luckily, we met some friends of Miss Smythe's, who invited us to tea. There is a Scotch Regiment out there, and the Y.M.C.A. have a tea hut, with a large tent erected, where the boys can get tea and cakes, etc, and sit down and play games, and listen to a gramo-

283. Now known as the Montaza Palace, built in 1892 in Alexandria
284. Rear-Admiral Horatio Nelson successfully led the British Fleet against the French Republican Army in the Battle of the Nile at Aboukir Bay in 1798

phone; otherwise, there would be nothing but sand, men, and sea. *En route* one passes sandy stretches and numbers of palm trees hanging with large bunches of ripe dates, at Montaga.

"We went to Cairo on Thursday, and stayed there all night, sleeping on the train. Miss Rundle and I went out to the Pyramids and had dinner with the Sisters at Mena House. We walked up to the Pyramids, which were bathed in moonlight. The evening sky was beautiful. We saw the Sphinx from a distance, and the native burial ground stood out clearly in the evening light. Byram—Egyptian Christmas—is just over, the natives dress festively, and go out to feast on the graves of their departed. The plain below the Pyramids is covered with water just now and looks beautiful.

"Some of the men tell me they pay awful prices on the Gallipolli Peninsula, for things. —4/ per dozen for eggs; 4/6 for 1 tin milk; 4/6 for 1 tin fruit, etc.[285] A canteen has been started, but as everyone is short of goods, hundreds are unable to buy there. The boys say it would be a great boon to get groceries sent out. I managed to get some cigarettes, tobacco, pipes, etc., through the Red Cross for wounded with some of the money sent by the Maitland Musical Society. I gave some of the things away to wounded on the train.

"Just now the Red Cross are giving to each man a square envelope, shaped calico pocket, containing pencil, paper, soap, washer, and a handkerchief and small bottle of eau-de-cologne. Wal is well. We see some of the 18th coming through on the hospital trains now and again and always enquire for Wal. I sent his last letter to Don. We are having nice weather now, though the sun is hot during the day if immediately in it. Rain is expected daily. Miss Merivale is sending me 12 dozen toothbrushes through Musical Society Fund, and I am going to give those out. Almost every man requires one, so they soon go.

285. According to the Australian War Memorial, the pay for a Private in the AIF was set at 6/- per day, suggesting the problems soldiers faced at Gallipoli

"What a beautiful memorial service that was. Bob would be proud if only he could have known how much he was thought of. Bob Kerr came to meet me one day last week at Cairo. We spent the time he had before catching his train in driving around Ghezireh. He looks well and expects to leave for the front some time this week. He is pleased to be getting away at last."

In October Louisa wrote to her older sister Mary[286] relating her work on the hospital trains, accompanying the wounded when they were transferred from the hospital ships. She noted that there were only about eight of the original No. 2 AGH staff still at Ghezireh.

"The others are on transport duty or on Lemnos or in England."

Some of the letter contains details of a parcel sent to Australia via Lieutenant Wells of the 2nd Battalion Gunnery section and how money collected by the Maitland Musical Society was being spent. There is also an account of an outing to Nousha Gardens[287] and reports of the weather.

"Since arriving in Egypt I have only known rain to fall three times, just for a few minutes ... the trees here are always covered in dust."

Alexandria, Egypt. 1915. Wounded soldiers from Gallipoli being transferred from a hospital ship to a hospital train for onward movement to Cairo. AWM, H12939

286. Letter dated 11.10.1915, collection of Dinah Toohey
287. Now known as the El-Nozha Gardens, they were made up of several gardens over 150 acres and were established by a wealthy Greek citizen, John Antioniadis, in 1860

In December, she was still resident at the Quarantine Depot in Alexandria with Miss Smythe. The *Mercury* reported that "nine cases of comforts (had been) forwarded from Maitland for the troops".[288] These were held in the British Red Cross shed in the docks. A further six cases were found and two more were missing. Louisa was taking steps to see these were distributed.

> "It is extremely kind of all to work so hard in the interests of the men," she wrote, "but they richly deserve all that can be done for them. I do not know if your committee is aware of the fact that under The Hague Convention, the Red Cross Society is debarred from providing the men in the fighting line with any goods whatsoever. This, if not known generally, will explain why men at the front do not receive a plentiful supply of shirts, socks, etc. Re extras in groceries, I have spoken to several of the men, who have been at Gallipoli. They say that parcels of that description would be most acceptable. Warm singlets, underpants, socks, etc., will be required for the severe winter weather. Handkerchiefs are always acceptable. Some of the men find that small bags of sulphur are of some use in keeping vermin away. A large number of these were being made in England, however, the winter weather should affect such pests."

She had been in contact with Private Brassell,[289] Robert Scobie's batman and also with his orderly, Private Richards[290] and both were "doing very well."

Later that month one recipient included mention of the 'comforts' in a letter to his mother[291] published by the *Mercury*. This was Private Wilfred Holden,[292] who had been a member of the Maitland Musical Society. He had been in hospital, he told his mother, and when Sister Stobo discovered he was from Maitland and a member of the musical society, she brought him 'comforts' purchased with money sent by the society some few months previously. He had been delighted to receive them.

In a letter of thanks to the Musical Society,[293] some months later, Private Holden mentioned part of Louisa's process.

288. 'Comforts for the Troops', Maitland Daily Mercury, 4 December 1915, p. 4
289. Possibly Francis Brassell, Private, later with the 1st Australian Machine Gun Battalion, formed in March 1918
290. Possibly Ernest John Richards, born 1890 and served with the 2nd Battalion. He was killed in action in France in April 1916
291. 'Hospital comforts', Maitland Daily Mercury, 23 December 1915, p. 4
292. Wilfred James Holden was killed in action in France in June 1918
293. 'Comforts for Maitlanders on Service', Maitland Daily Mercury, 22 February 1916, p. 4

"I was unfortunate enough to contract dysentery," he wrote, "and while in the Victoria Hospital at Alexandria, Sister Stobo visited there and enquired if there were any Maitland boys there. I had a long chat with her, and she mentioned that the society had asked her to spend certain moneys on its behalf. Well, I got some very useful and necessary articles and naturally feel indebted to the society and also to Sister Stobo for her kindness in looking out for other Maitlanders. I think we must consider ourselves very fortunate that we have someone in a position such as hers taking so keen and deep an interest in the welfare of the Maitland boys."

By the end of 1915, the Gallipoli campaign had reached a stalemate for both sides and finally, the AIF troops were evacuated from the Peninsula on 19-20 December.[294] For the Australians and New Zealanders, members of the AIF, the AMC and the AANS, the theatre of war was moving to the Western Front. This meant that the next part of Louisa's service saw her travelling between Egypt, Malta, England and France, and back again, with her attestation 'papers' briefly recording these moves. But Louisa's letters and the letters of others continued, which, at times, help to establish her whereabouts at more length.

One of the first mentions of her from 1916 came in a letter from Fred Phillips to his sister, Irene, who had returned to Australia.[295] Among other news, he told her that, "Sisters Stobo and Fillans got a trip to England." By this time, Irene Read had also become a fierce advocate for increasing 'comforts' to the troops. By the end of that month, however, it appears Louisa had returned. In a letter[296] from Corporal Alf Pender,[297] he states that Louisa was still in Alexandria at the time of his writing, as he speaks of trying to see her "as soon as possible."

By April, the *Mercury* was publishing more direct news[298] when Louisa wrote to the Mayor of Maitland, thanking him for £9/9/- sent for the troops and mentioned correspondence with the Musical Society. She had asked for their

294. Gallipoli, Australian War Memorial, https://www.awm.gov.au/articles/encyclopedia/gallipoli
295. Letter to his sister, Irene dated 3rd January 1916, Mitchell Library, MLMSS 2836 (K22194/Folder 8)
296. 'Corporal Alf Pender', The Maitland Daily Mercury, 18th March 1916, p. 11
297. Alfred William Pender, 13th Australian Infantry Battalion
298. 'Letter from Sister Stobo,' The Maitland Daily Mercury, 13th April 1916, p. 4

approval of her manner of spending some of their funds - £71/14/10 - for men in the trenches.

> "Those in the hospital get almost all they require while there," she wrote. In the letter, she makes mention of having been in England and in Malta, where she also established that those in hospital did not need extra comforts. She then encouraged Maitlanders to contact her directly regarding any articles they needed while they were in the trenches, using the address, c/o Australian P.O. Alexandria.
>
> "I only hope they take advantage of my offer," she wrote. "Just at present it is better to address parcels and letters with just the soldier's rank, name, number and battalion to Egypt as many battalions are split into halves and fresh battalions formed of each half, with reinforcements."

She elaborated on her efforts in the matter of 'comforts' in a letter sent on 22 April,[299] noting that she occasionally used other people to help distribute them in her absence, such as when travelling to England while serving on one of the hospital ships, in this case, the *Letitia*.

H.M.H.S "LETITIA" — of Anchor-Donaldson Line.

The HMHS *Letitia* was a British Hospital Ship

299. 'Australians in Egypt. Maitland Musical Society's Assistance. Nurse Stobo's Thanks.' Maitland Daily Mercury, 22nd April 1916

"No one knows what suffering those boys have gone through," she wrote. "Boys went through to England with us (English) with both feet amputated, owing to frostbite, and I was told that many died through the effects of exposure, etc. We can only hope and pray that the soldiers never again have to pass through such straits.

"I will spend the bulk of same should I go to England and bring the goods out with me," she continued. "I am enclosing herewith a summary of the money spent."

At times, she improvised in her efforts to assist as with a soldiers' lunch at Malta.[300]

"I was in a tea shop at Malta," she explained in her letter to the Musical Society. "A number of convalescent soldiers were having a light lunch. One was paying at the desk. Having seen him before, I went along and asked to be allowed to pay, as I had some money sent from Australia to spend on the soldiers."

On another occasion in Alexandria, she noticed a young soldier walking along the street.

"The boy was a West Australian," she wrote. "He looked lonely and 'hard up'. We passed him by and then I ran back to him and found he was just back to Alexandria from some outlying post. Upon enquiry I found he would be glad of some cash and that was all I had with me, (5/-) so I gave it to him, telling him where it came from."

In Malta, she bought cigarettes, which she took to England, asking her nephew, Don Scobie to distribute them to the men at Abbey Wood Convalescent Home.[301] She gave money to Sister Durham, who was accompanying soldiers travelling back to Australia on one of the hospital ships. Other articles were given to soldiers travelling from Alexandria to Cairo, from the hospital ships. She sent stationery to the camp at Tel-el-Kebir[302] via some Australian soldiers —various battalions—who were waiting on the Cairo platforms, as well as goods she had bought at Selfridges. She parcelled up pipe lighters —a dozen in

---

300. Due to its distance to the battlefront, Malta became a medical recovery outpost
301. 5th Convalescent Depot, at Borstal Heath, Abbey Wood, London
302. Tel-el-Kebir Camp was 40 km west of Ismailia and was a training centre for AIF reinforcements

each box—and wrote 'From the Maitland Musical Society' on each box. Her aim in these efforts was to get 'comforts' directly to the men.

Nurses with convalescent soldiers in Malta. AWM, H01615

Don Scobie wrote to the Society from the 5th Convalescent Depot at Abbey Wood, thanking them for the 'comforts' purchased with their money.

> "Cakes, cigarettes etc., were bought with the money," he wrote, "and I divided these out amongst the boys as best I could. I told each hut who the senders were as probably you will have some writing you. As far as I know I am the only Maitlander here at present, although there are ex-Maitlanders and a fair sprinkling of our 18th Battalion. Once more thanking you for your society's patriotism and enterprise, that has given us such a treat."

A 4th June letter[303] from one Maitlander, Corporal Jim Brassell[304] gives details of Louisa's departure from Egypt a week before. Brassell was in hospital with a fractured skull.

303. 'Local News', Maitland Weekly Mercury, 15 July 1916, p. 4
304. James Cornelius Brassell, 19th Australian Infantry Division. Born c.1885

"Maitland people do not know her worth," he wrote. "There is nothing too much trouble for her as regards Maitland boys. She even goes as far as putting up a notice in the homes for convalescent soldiers, asking Maitland boys to write to her and then she sends them money and comforts, the gifts of Maitland people. Her time is limited as she is on the hospital ship *Letitia* and does all the work in her spare time. She sailed for England last week and is going to look up the boys in France. It is hard after all the roughing we have done to finish up like this, but I am as good as ten dead ones yet."

As Kirsty Harris points out[305] nurses serving on ships shared the risks and conditions common to vessels in a war zone. This transport of wounded soldiers by ship was considered more dangerous than by road, train or other means, although hospital ships were theoretically protected by the Geneva Convention. They could be identified by prominent red crosses and a green band, painted across the white ship. Harris explains that most hospital ships were under the command of the British Royal Navy and that Australian nurses worked on troop transports and 'black ships' as well as hospital ships, adapting their practice to the ship's conditions.[306] Common procedures such as X-rays and operations were made more difficult by the weather and the often hot, dusty conditions. Even washing patients was harder when rough seas upset necessary equipment.

A letter from another nurse, Lindsay Gray dated 21st August[307] brought more news of Louisa. Nurse Gray wrote that she had spent time with Louisa in London and was happy to see her again. She said Louisa had newly arrived but had orders to leave their ship at Malta and join the hospital ship *Guildford Castle*.[308]

Louisa's attestation records show her as serving on the *Letitia*[309] as early as November 1915. The ship was attached to No.3 AGH, which was reestab-

305. Harris, K. (2009). Red Reflections on the Sea: Australian Army Nurses Serving at Sea in World War 1. The Journal of Australian Naval History, 6(2), 51–73. https://search-informit-org.ezproxy.lib.uts.edu.au/doi/10.3316/informit.261756961849214

306. Harris explains that 'black ships' were transports temporarily used to evacuate casualties from Gallipoli. Harris, Red Reflections on the Sea: Australian Army Nurses Serving at Sea in World War 1, p. 52-55

307. Letter from Nurse Lindsay Gray, Northern Star, Lismore, 4 December 1916, p. 6

308. The *Guildford Castle* was requisitioned as a hospital ship in September 1914

309. The *Letitia* was requisitioned as a hospital ship from November 1914 until she ran aground in August 1917

lished in Abbassia, early in 1916, staying there for a total of 10 months. Louisa had accompanied the sick and wounded from Egypt to England in May and returned in mid-June. By August, she had become a patient in the hospital herself. On 7th August,[310] the *Mercury* reported that she had been recovering there. Her attestation 'papers' show that she had mild tonsillitis on 26th July and was discharged for duty on 2nd August. On 13th August she was embarking on the *Kanowara*[311] from Malta to join the hospital ship, *Guildford Castle*. According to an article in the *Mercury*,[312] Louisa was expecting to go to Salonika, where small units of the AIF had been fighting in the Serbian Campaign. Kirsty Harris cites nurse Daisy Richmond as describing good conditions on board the *Guildford Castle,* with hot and cold water, sterilization equipment and numerous surgical instruments.[313]

In September, Louisa wrote to her sister May.[314] She had arrived in port that afternoon to disembark patients from Salonika. The wards had been packed with patients and the conditions had the ship pitching and rolling. She seems to have been on night duty and so had not been sleeping well. She was due to embark again soon and had no news of "the boys" as she called them, possibly referring to her nephews, Wal and Don.

> "We had some very sick people aboard," she wrote, "dysentery and malaria mainly with some wounded … We fully expected to be sent on to England upon our arrival here but heard we are to sail again tomorrow so I suppose that means Salonique again.
>
> "I have an awfully thick head tonight through want of sleep etc."

She goes on to speak of soldiers who had been on board, including a Captain Tennant who had crossed the Atlantic 20 times.

> "He got into the way of greeting me as Advance Australia!" she wrote. "… There is no fresh news so I will close to get this off by the mail. Goodbye for the present. With love to all. Your affect. (affectionate) Louisa Stobo."

310. 'Illness of Sister Stobo', Maitland Daily Mercury, 7 August 1916, p.4
311. The *Kanowara* was requisitioned as a troop transport ship but in 1915 it was converted into a hospital ship, transporting sick and wounded Anzacs between Australia, England and Egypt
312. 'Assisting Maitland Men', Maitland Daily Mercury, 13 October 1916, p. 4
313. Harris, K. (2009). Red Reflections on the Sea: Australian Army Nurses Serving at Sea in World War 1. The Journal of Australian Naval History, 6(2), 51–73.https://search-informit-org.ezproxy.lib.uts.edu.au/doi/10.3316/informit.261756961849214, p. 57
314. Letter to May, 29 September 1916, Scobie Family History

Louisa's 'papers' show that on 14th October 1916 she was to report to No. 3 Australian Auxiliary Hospital (No. 3 AAH)[315] for duty at Dartford England and by 1 December, the *Mercury* was giving details of this posting.[316] They wrote that she was nursing sick and wounded soldiers in Dartford and had previously made three trips to Salonika on the *Guildford Castle*. On her last return to Malta, she had been given less than an hour to prepare for embarkation on the *Gloucester Castle*, for passage to England, noting that they were leaving the sunshine and warmth behind in exchange for cool weather and grey skies. Five months later, the ship was torpedoed by a German U-boat.[317]

Louisa, at left, with two other nurses.[318]

Along with nursing, Louisa continued her work on behalf of Maitland soldiers. A letter from Bombardier Storm[319] to the President of the Maitland Women's Patriotic Committee, mentions her efforts.

> "I have the pleasure of thanking you, on behalf of my comrades for the money handed to me this morning by Sister Stobo", he wrote. "I am sure the boys will be exceedingly pleased, and appreciate the money like myself, as you know we have had no pay for two months while in hospital."

315. The Orchards Convalescent Hospital was on the riverbank at Joyce Green Lane Dartford, Kent, before it was handed over to the military and became No. 3 Australian Auxiliary Hospital on 9 October 1916
316. News of Sister Stobo, Maitland Daily Mercury, 1 December 1916, p.4
317. The *Gloucester Castle* was torpedoed by a German U-boat on 30 March 1917
318. Collection of Dinah Toohey
319. 'Patriotic Money, Letter of Appreciation', Northern Times, 11 December 1916, p. 6.

Louisa on duty at No 3. AAH[320]

In late October 1916, Louisa wrote to her sister, Janet, describing No. 3 AAH.[321] In it she gives news of Maitland boys, including her nephews, Wal and Don. There are glimpses of her life away from the hospital and also the likelihood of her remaining in Europe even if she is given the opportunity of returning.

> "You will be wondering if I am still here," she wrote. "It will be a week tomorrow since I arrived. This place is built in pavilions, each containing 46 beds. The wards are comfortably furnished with hot water pipes on either side for heating purposes. There must be about, or over, 1000 patients here now. They are being drafted from other hospitals and are sent from here either on furlough to Weymouth for further examination or are boarded for return to Australia. Our cubicles are very comfortable too. Each sister has one with dressing table combination, wardrobe, etc, and hot water system therein."

320. Collection of Dinah Toohey
321. 'Some of the patients. Letter from Nurse Stobo'. Maitland Daily Mercury, 23 December 1916, p. 9. No. 3 AAH specialised in treating war-related nerves

The letter speaks of her contact with Lieutenant (now Captain) Sawyer's family, whom she had evidently visited before, and with relatives of the Stobo family in England.

No. 3 Australian Auxiliary Hospital, Dartford, England. AWM, P00162.020

"I have been to London three times since coming here," she wrote. "The first time visited the office and did some necessary shopping with Sister Gordon. The second time, last Sunday afternoon, I went to visit Capt. Sawyer's people. They are so homely and nice and were so pleased to see me again. The following day I went into town to meet Tom Stobo,[322] who was staying at the Y.M.C.A. Aldwych huts in the Strand. We went out to see the Elsmies.[323] One of Mrs. Elsmie's sons is a Brigadier-General and is wounded. Another son is a Colonel in India. In fact, the whole connection seems to be more or less military. Mrs. Elsmie has had a bad attack of jaundice lately. She is a fine-looking, handsome woman. Aunt Bella was also there, and a daughter, Mrs Hamilton from India.

322. Thomas Charles Wren Stobo was the son of Thomas Stobo, Edye Stobo's brother
323. The family of Elizabeth Spears Elsmie, whose sister was Mary Dunlop (Spears) Stobo. One son was a Major-General of the British Indian Army

"Fred Phillips has gained the soldiers' special war medal and is now a lieutenant in the 2nd Battalion machine gun section.[324] He is very pleased that he got the medal before he got his commission. Sergeant Jim Watson[325] (Bob Watson's son) is in this ward. There is a boy named ... a young friend of Wal's, in another ward, and young Storm was in another. All are progressing and are able to go out."

Her letter relates that physical conditions for nurses in England were more comfortable, but they were shocked by the news of casualties in France.

"We have had a couple of white frosts since coming here, also rain and fog, but on the whole, we do not seem to mind such particularly for we can get piping hot baths when off duty. Burrows, Welcome and Co.[326] have a large factory near the railway station at Dartford. I have written three letters to Jim Kerr[327] and have not seen or heard of the boys except through some ancient letters to Egypt. I am forwarding any letters I receive from Wal to Mick, as there may be some news in them. He was still at Cayeux[328] according to head office reports and seemed to be bright and having a fairly good time. He was on light duty.

"From all accounts the fighting in France is fearful, and the casualties dreadful. There is a flying school near here and the machines are buzzing overhead all day, and at night the search lights are playing from all quarters. Vicker's factory[329] is not far off, and if an air attack is expected the lights are extinguished, otherwise it is a blaze of light. We are taken to and from the station by motor ambulance and return here usually by 9.30 p.m."

She concludes the letter with news of her wounded nephew's return to Australia and the likelihood or not of herself returning to Australia in the short term.

324. Frederick Stobo Phillips was awarded a Military Cross for conspicuous bravery at Pozieres as a stretcher bearer
325. James Watson, 17th Battalion
326. Burroughs, Wellcome and Co. manufactured products for the prevention and treatment of disease
327. Louisa's cousin, James Kerr, 53rd Infantry Battalion, born Maitland in 1885
328. A seaside town c.30 km northwest of Abbeville, which is located largely on the east bank of the Somme River
329. A munitions factory in Crayford, Kent

"Don will probably be home ere this. There is some talk of sisters who are from Australia being given the opportunity of returning, I do not know if I am likely to avail myself of this opportunity. I had quite a large budget of letters to read on my arrival in England, but they were almost all written long ago. Perhaps the next budget will be now. Sister Gordon has been told to prepare for transport duty early in November."

More news came in a letter of 6th December.[330]

She wrote that she,

"fully appreciated the contents, especially the gowns. They are some 'knut'[331] and are greatly admired. I have distributed the sweets, etc., amongst my pals. The boys got a few sweets, too. It was awfully good of them to send the things. Excitement is great when parcels arrive."

In a letter of 14th December[332] Louisa wrote that she had received a parcel from home that week. Two cases of 'comforts' had arrived from 'The Australian Nurses' Comforts' Fund', which Louisa was to distribute. Some of these were ticketed from the West Maitland Women's Patriotic Fund and Louisa and Sister Gordon spent an afternoon sorting and distributing a parcel to each sister on staff.

"You should have heard the exclamations of pleasure and surprise," she wrote, "when they saw the parcels on their beds and listened to the sisters when they came off duty — 'What did you get?' 'Just what I wanted,' etc. I managed to supply 44 persons with gifts and gave the groceries to the sisters' mess."

The new year found her still at No. 3 AAH, and still providing 'comforts' where she could, as a letter[333] from Private Frank Flannery[334] shows.

"I am in receipt of the sum of 10/- from Sister Stobo of this hospital, which I understand is a gift from the Maitland people and beg to tender my very sincere thanks for same", he wrote. "The gift

330. 'Maitland Soldiers, Letters from a Nurse,' Maitland Weekly Mercury, 10 February 1917, p. 7
331. This was a jocular term of the era and referred to the idea of fashionableness or dandyism 332
'Nurses' Comforts,' Maitland Daily Mercury, 1 February 1917, p. 4
333. 'A Soldier's Appreciation,' Maitland Daily Mercury, 27 January 1917
334. Francis Victor Flannery was 18 when he enlisted in August 1915

came just at the time when I needed it. I had not received any money for about three months, as it is not the custom over here to pay men in hospital. I managed to get lots of little comforts with the money which otherwise I should have gone without."

More news came in February of Maitland soldiers.[335] Louisa had finally met Jim Kerr, who looked well but thin, while her nephew, Wal Scobie, was once again ill and at No. 39 Hospital in France. She mentioned the names of other local boys, including Dick Long, Jack Power and Bob Kerr. Sergeant Watson was in Weymouth and Private Flannery was helping out at the hospital at the orderly office.

"Lots of the boys are having furlough at present," she continued. "One boy told me that Capt. Gordon Chant and other officers were having something to eat in a dug-out when a shell lobbed into it. If so, their end would mercifully have been painless."

The weather had been more settled, she continued, but the autumn had been the wettest in 60 odd years.

"Some of the patients have trench feet," she wrote. "The feet become bruised looking, tender, and swollen. There are dozens of maimed boys here and they are so bright. Those who are suffering from shell shock and the after effects of being buried are most dreadfully nervous as a rule. Several came in having lost their voices, but they usually get over it after treatment. We have some awfully nice boys in our ward now. There is a cousin of Mrs. Bob Watson in the casualty room — a Mr. McGrath, from Ballarat.

Louisa continued, saying she had met soldiers who, as yet, had not benefited from the Maitland 'comforts' funds.

"I have met several men of the old 2nd Battalion", she wrote, "and not one has had any benefit as far as he knows, from that fund. I particularly enquire and each tells the same tale and says no such assistance or benefit has reached them, not even a canteen for their benefit. I have thought several times of asking the 2nd O.C.[336] to send me a sum so that I could hand some of it to 2nd and 54th

335. 'Maitland Soldiers, Letters from a Nurse,' Maitland Weekly Mercury, 10 February 1917, p. 7
336. Officer in Command

men in hospital, etc., for it is a pity to let the money be idle if such is really the case."

A week later she wrote again with news of a young soldier. Her interest and compassion are evident in the details she relates.

"We have a boy named Howell in our ward who is related to the Thursbys.[337] He is only 15. He has lost three fingers, portion of the forefinger, has a wound on the wrist, which is stiff, a stiff right elbow, necrosed bone in his foot, and his front teeth have been extracted. The boys call him 'Fragments'. They tease him for being young, but he says, 'Anyway, I am the making of a man'. A man named Reed has chummed up with him and they vie with one another for the Dartford girls' smiles. One of them is on crutches, the other with a stick. I told Howell he was to go to Maitland when he returned to Australia. He is quite a nice youngster.

The letter continued with news of the Western Front.

"The weather is chilly here," she wrote. "Some of the boys are on furlough from France. They say the mud is simply awful, up to their waists in places. They almost all wear sheepskin vests over their tunics and find them very warm, especially as they are unable to have their blankets with them often times. One boy told me he had not seen a blanket for weeks."

Members of the 1st ANZACS corps cyclist battalion, wearing sheepskin vests.
AWM, P10550.160

337. Frederick Thursby was Mayor of West Maitland in 1911

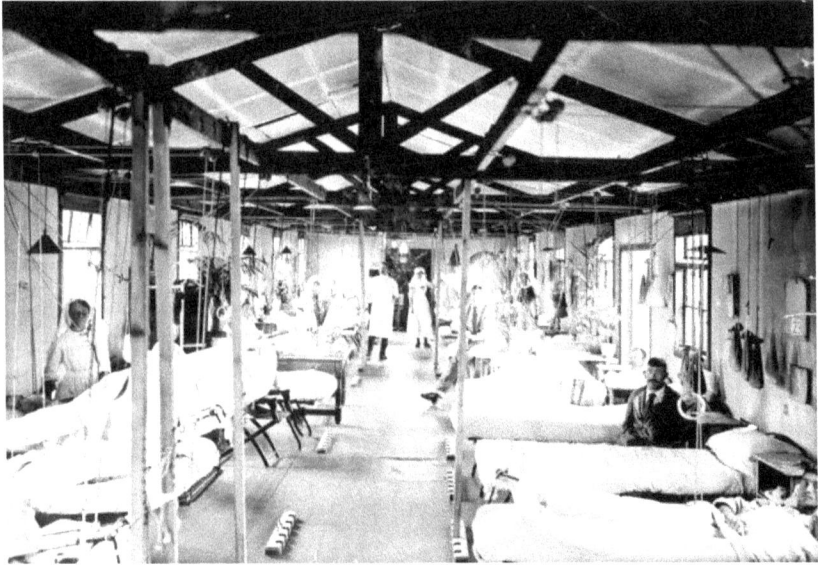

Inside a ward at No. 2 Australian General Hospital, Wimereux. AWM, PO1630.002

These intimations of conditions in France are some of the first impressions we have of Louisa's next billet. By March 1917,[338] the *Mercury* was reporting that Sister Stobo, who had recently been discharging the duty of night superintendent at 3 AAH, Dartford, had received orders to leave for No. 2 AGH at Wimereux, c. 7 km from Boulogne-Sur-Mer, in France.[339] This was the original hospital unit she travelled with on the *Kyarra* and where she had served her first several months in Cairo. Sister Agnes Shearer was also posted to No. 2 AGH and has left this description.[340]

> "It was a large tented hospital, and we were billeted in most uncomfortable tents. It was a very severe winter, and we had no heating apparatus, and the tents were not weatherproof. Afterwards, however, conditions improved, and huts were erected and we had stoves in nearly all the cubicles."

In an undated letter to her mother from France, Louisa was quick to offer reassurances that there was no need for anyone to worry about her as they

---

338. 'Sister Stobo in France,' Maitland Daily Mercury, 17 March 1917, p. 4
339. A large, tented hospital for battle casualties
340. Agnes Shearer, AWM 41: 1072. Interviews containing accounts of Nursing experiences in the AANS [Australian Army Nursing Service]

were "miles and miles away from the fighting."[341] She reported that they were not yet in their own hospital but rather stationed at a Royal Ambulance Hospital that had been there for two years. It was on the same hill as No. 3 AGH and another South African hospital and had extensive and pretty agricultural views. They were sleeping in tents and had been busy with wounded from English and Scottish regiments.

No 2. AGH was at Boulogne, she noted, and she and other Sisters were having trouble adjusting from the snows of London to the warmer weather of France.

A dugout for nurses of No. 2 Australian General Hospital,
Wimereux, France,1916. AWM, H04152

"Boulogne is interesting," she noted, "but grimy looking and dusty." Louisa had written that she had received two sums of money - £50 and £13 for the relief of soldiers but was unclear as to who had sent it. She believed it had come from Maitland but was awaiting further advice. In the meantime, she intended to spend some of the money on socks to help protect the men from the mud and wet weather in France. The last 10/- would go to her young patient, 'Fragments' Howell.

The *Mercury* made mention of Lieutenant John Saunders,[342] who had been at the Somme and described the holes left by shelling; some were 10 feet deep,

341. Letter to Mary Scobie, addressed No. 3 AGH, C/o 2nd Stationary Hosp. B.E.F. France. Collection of Dinah Toohey.
342. John Joshua Saunders, 14th Machine Gun Company

and they needed the help of horses when men fell in. Louisa supplied Saunders with money and groceries for men in the trenches which he would distribute on his return to France, as he saw fit.[343] Saunders subsequently wrote to say how much the men in the trenches had appreciated the 'comforts'.

> "Of course, fires were out of the question", he wrote, "but they had 'fixed' the gun teams up with primus stoves, so they were able to drink cocoa and milk etc. Such gifts coming at a time when they went in the trenches made their work a little easier."

In early March, the Women's Patriotic Committee of West Maitland were active on her behalf as the *Mercury* reported.[344]

> "Sister Stobo ...", they reported, "kindly undertook the extra duty of distributing the money from Maitland. She also forwards money to nurses of hospitals wherever our boys are located. The soldiers in hospitals receive a reduced pay and often need comforts and necessaries that they cannot procure for want of funds."

Louisa's 'papers' show her as attached to No. 2 AAH Southall on 12th March 1917, and detached to join No. 3 AAH on 16th March. No. 2 AAH was located in the St Marylebone School in Southall, Middlesex. The Virtual War Memorial[345] states that it was staffed by personnel from three auxiliary hospitals, the 1st, 2nd and 3rd. Its speciality was fitting artificial limbs. The next day, 17th March, she was detached from attendant duty to Headquarters, at Horseferry Road, London. According to the Australian War Memorial,[346] these were small military headquarters, set up to provision the AIF and to manage paperwork attached to AIF convalescents and medical unit personnel.

On 3rd May her 'papers' show she "proceeded overseas" to join the British Expeditionary Forces and a week later they show her as transferring from No. 3 AGH, Abbeville to No. 2 ASH. These stationary hospitals were set up in more forward areas than the general hospitals. No. 3 AGH operated from

343. 'Comforts for Soldiers,' Maitland Daily Mercury, 14 May 1917, p. 4
344. Women's Patriotic Committee, A Gift Evening, Maitland Daily Mercury, 6 March 1917, p. 4
345. 2nd Australian Auxiliary Hospital – WW1, (n.d.), https://vwma.org.au/explore/units/1341
346. Australian Imperial Force Headquarters (London), Administrative registry medical/personal files, 1914-18 War, Australian War Memorial, (2024) https://www.awm.gov.au/collection/C2130826

tents and huts at Abbeville,[347] in the Somme and treated casualties briefly before sending them to other places. No. 2 ASH would have been a smaller hospital, located closer to the fighting.

A section of No. 3 Australian General Hospital, Abbeville. AWM, C04539

As she worked closer to the battlegrounds, there was less mention of her in local newspapers. However, in August, the *Sydney Mail*, in reporting the Anglo-French advance in Flanders, gave some news of her.[348] The *Mail* reported that the advance had been impeded by torrential rain but that thousands of prisoners had been taken. They quoted a letter just received from Sister Stobo, who they described as an Australian attached to one of the field hospitals in Northern France.

> "We had a violent thunderstorm during the night (early in June)", she had written. "Two guards were struck by lightning, and their two prisoners killed. About a week ago three orderlies were struck by lightning. One was marked from his right shoulder to his foot, and across the back; the other two were stunned for a time. Shortly before that two men were killed by lightning in the village. The rain pours in a deluge. Almost all of us had to move our beds, as they were getting wet through, owing to the roofs leaking."

347. Australian World War 1 Hospitals, Anzac Day Commemoration Committee, (2024), https://anzacday.org.au/australian-ww1-hospitals
348. 'Brilliant Anglo-French Advance in Flanders', Sydney Mail, 8 August 1917, p. 18

No. 2 Australian Casualty Clearing Station tent lines in Blendecques,
c. 12 kilometres from the town of St. Omer. AWM, H01993

Louisa was still corresponding with her family when she could. Her 'papers'
show her as attached for duty at No. 2 Australian Casualty Clearing Station
(No. 2 ACCS) at Trois Arbres,[349] in mid-July, with the *Mercury* reporting a
letter from her, written soon after.[350] In it, Louisa said she had been
transferred from No. 3 AGH to No. 2 ACCS, which was nearer the firing line
than any other. According to Lieutenant Colonel H. S. Stacey, who set up
No. 2 ACCS, it was 7000 yards from the front-line trenches at
Plogsteert. Along with No. 1 ACCS, it was "almost literally in the front
line".[351] The third Battle of Ypres was to start on 31st August and an influx
of wounded was expected. The Matron-in-Chief, Evelyn Conyers, had
prepared lists of nurses who she deemed suitable for duty in Front Areas
and Louisa was evidently on these.

"The patients were many, and the wounds in many cases were severe", Louisa
wrote in her letter.

> "The surgeons are kept busy operating almost all night and day.[352]
> We can hear the guns at the hospital, and on clear nights the aero-

---

349. Trois Arbres is c. 3km from the town of Steenwerck and a similar distance to the French/
Belgian border
350. 'Sister Stobo in France,' Maitland Daily Mercury, 19 September 1917, p. 4
351. No. 2 Australian Casualty Clearing Station, Through These Lines, (2012), https://
throughtheselines.com.au/research/2-ACCS#location-1
352. Trois Arbres, Through These Lines, (2024), states that c. 2000 operations were performed in
July and August, https://throughtheselines.com.au/research/trois-arbres

planes are continually overhead. It is simply wonderful to see them. When the enemy planes appear, the anti-aircraft guns sound forth from all points, and the searchlights flash across the sky. The machines, when located, look like giant birds, with outstretched wings."

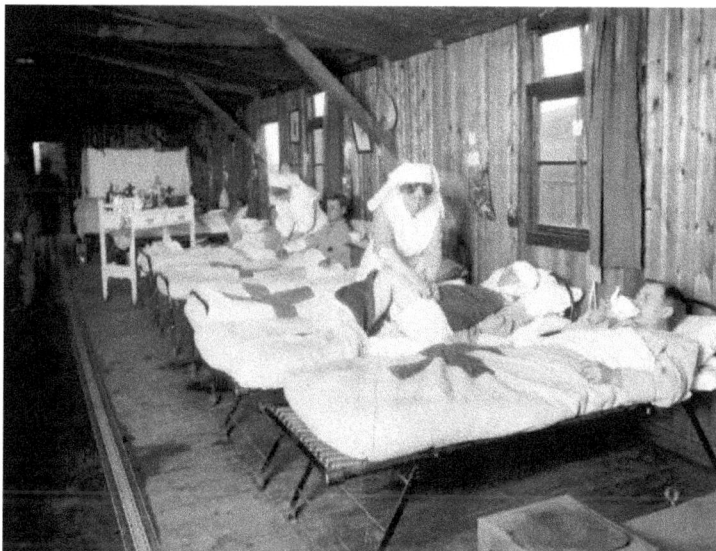

A ward in No. 2 Australian Casualty Clearing Station. AWM, E04623

Louisa had written that the nights were clear, and the moonlight allowed for air work, with the planes crossing and re-crossing continually. Three days before, on 14th July, bombs had been dropped in a potato patch about a minute's walk from the hospital. On the night she arrived, a bullet had passed through the ceiling of one of the sister's rooms and landed on the dressing table.

In her letter, Louisa mentioned that Major Faulkner, a Victorian, was the O.C. while she was sister-in-charge of a staff of ten nurses. Through These Lines[353] lists a total of three sisters-in-charge at No. 2 ACCS, Sisters Pocock,[354] Stobo and Davidson and a total of 19 other sisters. These additional numbers may refer to nurses who had been serving at No. 2 ACCS since its inception in July 1916.

353. No. 2 Australian Casualty Clearing Station, Through These Lines, (2012), https://throughtheselines.com.au/research/2-ACCS#location-1
354. Mary Ann (Bessie) Pocock was born in Dalby Queensland in 1863.

At one stage, Sister Agnes Shearer was also posted to No. 2 ACCS and has left a description of the set-up.[355]

> "It was a large hut and tent hospital", she recalled. "The conditions were very good, but we still had no heating. We got fresh vegetables, fruit and milk from the farms around."

Casualty Clearing Stations (CCSs) were the second stage of treatment for wounded men, after immediate treatment at dressing stations. CCSs were located on railway lines and the 2nd ACCS was at the rail head at Trois Arbres. Canvas wards and operating theatres were housed in tents nearby. One nurse who had worked at No. 2 ACCS in May wrote that they were receiving the men approximately an hour after they had been wounded.[356]

Louisa arrived in mid-July and wrote to her family on 17th July. Five days later, on 22nd July 1917, No. 2 ACCS was bombed at 10.25 p.m. by German aircraft, performing raids over Allied positions.[357] Two bombs were dropped. The first hit the pneumonia ward and blew a hole in the ground, killing two patients and two orderlies. The second dropped outside the ACCS boundary, wounding more patients and staff. Lieutenant Colonel J. Ramsey Webb recorded that four people were killed and 19 people wounded.[358]

Duckboard walkways allowed orderlies to transport patients at No. 2 Australian Casualty Clearing Station. A two-tiered stretcher holder on wheels allowed patients to be transported to waiting trains. AWM POO156.061

355. Agnes Shearer, AWM 41: 1072. Interviews containing accounts of Nursing experiences in the AANS [Australian Army Nursing Service], p. 107
356. Trois Arbres, Through These Lines, (2012), https://throughtheselines.com.au/research/trois-arbres
357. ibid
358. ibid

Four of the nurses - Claire Deacon,[359] Dorothy Cawood,[360] Mary Jane Derrer[361] and Alice Ross-King[362] – showed such gallantry, they were awarded the Military Medal, the nurses' equivalent of the Victoria Cross. As with Louisa, these women had enlisted at the beginning of the war and served in Egypt before being transferred to France in 1916.

Alice Ross-King had been called to the pneumonia ward when the first bomb landed in front of her. She fell into the hole it created but managed to get out and keep moving towards the patients.[363] The orderly who was leading the way was blown apart. The other three nurses ran to the wards and tried to rescue patients or protect them as best they could. All ignored the men's cries to find shelter for themselves in a dugout.

For her part, Louisa was to receive the Royal Red Cross, 1st Class, which was awarded for "exceptional services in military nursing."[364]

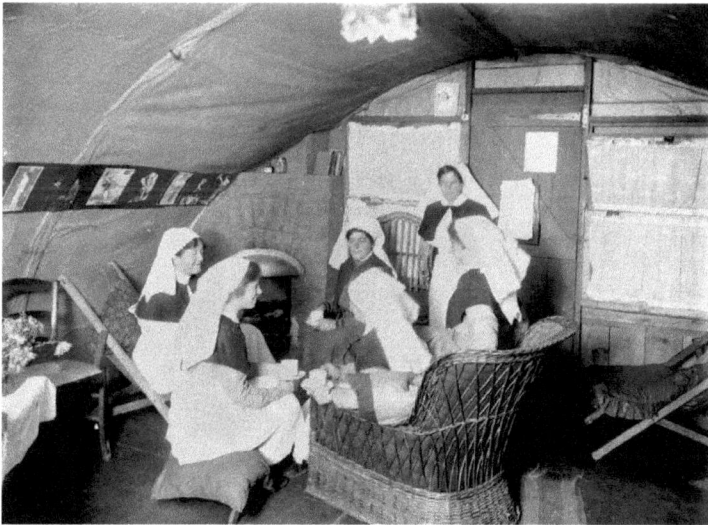

Nurses having tea at No. 2 ACCS. Left to right: Sister Mildred Crocker-Brown; Sister (Topsy) Tyson; Sister Helen Homan (centre background); Sister Louisa Stobo (centre foreground); Sister Steward (standing); Sister Shepherd. AWM, E01 280

359. Claire Deacon was born 1890 in Crows Nest NSW
360. Dorothy Cawood was born 1884 in Paramatta NSW
361. Mary Jane Derrer was born 1892 in Mackay Queensland
362. Alice (Alys) Ross King was born 1887 in Ballarat Victoria, had sailed on the *Kyarra* and served at No. 1 AGH in Egypt. She was promoted to Head Sister when the hospital moved to Rouen, France
363. Stewart, Elizabeth, (2021), Nurses Under Fire, Australian War Memorial, https://www.awm.gov.au/wartime/50/stewart_nurse
364. Lemnos Gallipoli Commemorative Committee Inc, (n.d.) https://lemnosgallipolicc.blogspot.com/2014/12/royal-red-cross-medal-australian-ww1.html#:~:text=To%20recognize%20further%20exceptional%20devotion,is%20made%20of%20red%20enamel

Many of the patients they received were victims of mustard gas attacks. From her 'papers', Louisa remained as Head Sister with the unit until mid-February 1918; other reports state she remained in the role until late-February.[365] This means she was the Sister-in-Charge referred to in Through These Lines who wrote "Mustard oil shells are being used by the enemy, in consequence of which we receive many patients with burns therefrom, the eyes specially being much inflamed. At times, large blisters form on the body.[366] No. 2 ACCS was shelled again in September 1917 and the operating table bent by a shell that came through the roof of the hut.[367] No one was injured but casualties remained heavy with 2-3 surgical teams working through the night. Sister Derrer reported that there were eight operating theatres working day and night and the wounded were "as bad as could be… If one had time to think we would have just been weeping hysterical women but we'd only time to do."[368] In November, one of their patients was Private John Moylan of West Maitland, aged 18. Moylan had returned to the line after carrying out a wounded comrade when a shell burst near his dugout. One of his legs was severed and he died of his wounds. In a letter of condolence to his family, Louisa included details of his final hours.

> "I am sorry to have to tell you of the death of your son," she wrote. "No doubt you will be pleased to know there was someone here who knew of his people. The boy was admitted to the 2nd ACCS Hospital in France on the evening of the 20th November and died the following morning. As soon as I heard the name I remarked that perhaps he was a Maitlander, so went to the ward, and found such was the case. He was quite conscious, and I think was too much shocked to feel much pain. One leg was blown off, and the other badly injured. I told him who I was, and he said, 'I think I've got an Australian Sister'.
>
> "I asked him if he would like to see the padre", she continued, "and Chaplain Major Le Maitre,[369] who belongs to Adelaide, and is

365. No 2. Australian Casualty Clearing Station, (2012), Through These Lines, https://throughtheselines.com.au/research/2-ACCS#nurses
366. ibid
368. Sister Mary Jane Derrer, Australian War Memorial, http://www.awm.gov.au/education/resources/nurses
369. Edward Le Maitre, Australian Army Chaplain was born in Belgium in 1867 but resident in Glen Osmond South Australia

attached to this unit, visited him. The boy is buried in Trois Arbres cemetery, adjoining this hospital. It is a mile from Steenwerck and three hours from Calais. There are many of our home boys buried there. The boy was taken to the operating theatre on the night of November 20, was conscious after being fixed up but terribly restless and died towards morning. Hoping that this may be of some assistance to you in your sorrow, I remain, yours sincerely, Louisa Stobo, Sister in charge."

Le Maitre spoke of Australians he had attended when he returned to South Australia in 1918,[370] of their love of Australia and how soldiers shared letters from home with mates. He believed German soldiers both feared and respected the Australians because the latter were fierce fighters but showed kindness when taking German prisoners.

In the midst of these tragedies, Louisa continued her advocacy for soldiers 'comforts'. Private Holmes of Wallarobba had received £1/10/- from the Maitland Patriotic Fund via Louisa.[371] Other Maitland institutions, like the Horseshoe Bend Infants School raised £12/2/6, of which the greater part would be sent to the Sister Stobo Fund, while the Horseshoe Bend Patriotic League was sending a further £9/9/3 which they had raised through holding euchre parties.[372] Even some of the beneficiaries of the 'comforts' paid it back in kind, like Private G. Holmes, who sent 10/- to the Sister Stobo Fund.[373]

From 16th February 1918, Louisa's 'papers' show she was at the Nurses' Home at Abbeville until 12th March but during this period she travelled to London. There, on 20th February, King George V pinned the decoration of Royal Red Cross, 1st Class, on her at a ceremony at Buckingham Palace. Recipients of this and other decorations were then entertained by Queen Alexandra at Marlborough House.

This respite was brief and on 13th March Louisa is listed as reporting for duty at No. 3 AGH at Abbeville. As the fighting intensified in this phase of the war, her next move was on 31st March to No. 25 British General Hospital at Hardelot Boulogne for temporary duty. According to Through These Lines, the hospital was staffed by Australian Nurses and English Medical Officers and most of the hospital was under canvas.[374]

370. Le Maitre, Edward, Virtual War Memorial, (n.d.) https://vwma.org.au/explore/people/188137
371. 'Grateful for financial assistance,' Maitland Daily Mercury, 7 November 1917, p. 4
372. 'Sister Stobo Fund,' Maitland Daily Mercury, 17 December 1917, p. 4
373. 'Sister Stobo Fund,' Maitland Daily Mercury, 18 January 1918, p. 4
374. Through These Lines, https://throughtheselines.com.au/research/hardelot-plage

A long letter to her sister, May, dated 2nd April, gives details of Louisa's time there.[375]

"I am writing to tell you that the notification of that money you sent … has arrived. It is good of everyone to send money to help in this way, though just now I am at a very quiet and beautiful spot on the seashore near Boulogne, lovely sandy beaches and woods with primroses, cowslips, wood anemones and daffodils in bloom. The hillsides near sheltered parts are a picture. The trees are just budding in places, at others bursting into leaf.

"We were sent here from No. 3 AGH, really with the idea I should say of getting as many as possible down from the danger zone, in case things became more serious, though I feel things will straighten out again soon. The first two days, the Channel … prevented the arrival of reinforcements. It seems our luck to have the weather against us whenever a big move is on.

"We arrived here on Saturday. Some of the girls waited on the platform for six hours. Many have lost their entire luggage—glad to escape with their lives. The poor wounded boys had a dreadful time at first. One poor boy who was wounded abdominally hung onto the back of an ambulance though the sisters said he was very badly wounded. Others got out on the gun carriage, on tanks, etc., and I am afraid many fell into the hands of the enemy.

"I gave 20 francs each from your amount (used my own so far) to two patients I had; (Tommies) one Scotch, the other a Londoner, 19 years of age. The former lost both of his legs half way up the thigh, a married man too. The boy lost one arm above the elbow and the other arm below the elbow and had dreadful wounds in both legs. He is a grand kiddie. We had to give him chloroform to remove the dressing from his legs, as they were evidently, during the rush, put on dry on the raw incisions and had stuck fast. He told sister he had a lot to be thankful for, as he had no one depending upon him. I gave each one your address.

"One day, I had invested in 19 francs worth of cigarettes and saw one of our 11th Division boys, who had just come down from the

375. 'Nurse Stobo in France, Fleeing from the Enemy', Maitland Daily Mercury, 14 June 1918, p. 3

front, going along the road ahead of us. We hustled, and I gave him the package of cigarettes for himself and mates. I did not ask his name, simply told him the things came from West Maitland people, so if there is no acknowledgment of some of the things you will understand. Of course one gives out cigarettes, etc., in small packages to the patients. Apple-women sit about, selling fruit and chocolates to the patients here.

"This is a sea-side resort. During the season in peace time, princesses, duchesses, and other such people come here. The sisters are billeted in very nice houses. They have four or five of such. The patients all have skin diseases, boils, eczema, etc. It is a lovely place for them, for the air is fresh, and the sandy sea beach is beautiful. Miss Kellett, who was at the Sydney Hospital, is matron here with an Australian nursing staff. A train (one hour's ride) runs into Boulogne. We are 12 miles out by road."

By mid-to-late May, Louisa's 'papers' show her as posted to No. 2 AGH, where she remained until late September, when she was given two weeks leave to go to England. She reported back for duty on 13th October, still to No. 2 AGH. By 18th November she was sick with influenza and admitted to the 14th General Hospital, where she remained for ten days and by the beginning of December she was back at No. 2 AGH again.

Louisa turned 44 on 16th January 1919, with her 'papers' showing she was proceeding to Headquarters in London, preparatory to returning to Australia. Two days later, they show her as attached for duty at No. 2 AAH, Southall.

Her departure for home was delayed, however, by her completion of training as an Inspector of Nuisances at the Royal Sanitary Institute, 90 Buckingham Palace Road, London. She was granted leave from 24th February until 31st May, with subsistence pay of 6/- per day and a further £1/10/6 for lecture fees.

In April she wrote to her older sister, Elizabeth Roberts (née Scobie).[376] Louisa had received another parcel from home, with some hand-knitted comforters, in the socks and mittens. She had given the mittens to the Curator at the Institute, "an Irishman who is as thin as a lathe

376. Letter dated 4th April 1919, collection of Mary Thom

and had chilblains on his hands." The socks she intended to give to an odd-job man at the Institute. He was West Indian and had "got religion" as she wrote.

"I wish I had religion anyway," she continued, giving some indication of her state of mind after four long years of war.

A June letter to her sister-in-law, Flora, Robert Scobie's widow, gives more detail.[377] She wrote that she had passed the exam for sanitary inspector at the Royal Sanitary Institute, which had been held at Cardiff, in Wales. Something of the energy and determination that had seen her through five years of war can be glimpsed in this letter and the amount of activity she had packed into a short time.

> "The institute holds exams at various places", she wrote, "this time at Cardiff. So, to Cardiff we went. During the time we waited for the results to come out we crossed by boat to Weston-super-Mare, went to Bristol by train, stayed there on Sunday (one hour's (trip), and returned to Weston on Monday morning, whence we journeyed by charabanc to Glastonbury Abbey, Wells Cathedral, and Chedday[378] (sic) Caves. On Tuesday we went to Minehead and back, then on to Cardiff on Wednesday by train, and visited the show there before joining our train for London.
>
> "The country everywhere was just beautiful, field, hedges, and trees, from London, thence and back again. The Abbey and Cathedral were just beautiful. The Chedday Caves are very nice, but not nearly as fine as the Jenolan Caves. You should have seen the animals at the Cardiff show. There were wonderful Shorthorns, and varieties of sheep I had never seen before, excepting as wooden toys for children — absolutely square things. The horses were a picture to behold. Bristol is absolutely fine. We used sometimes to go to Battersea and Hyde Park to study and ask one another questions. The parks are a picture. I expect to be leaving here sometime in August, though one never knows for the sisters and others are held up for weeks and weeks waiting for boats."

377. Nurse Stobo RRC, Maitland Daily Mercury, 29 August 1919, p. 4
378. Louisa is referring to the Cheddar Caves, located near the village of Cheddar in Somerset, England

She returned for duty at the completion of the course and spent her last weeks of service at No. 3 AAH, Dartford. Finally, on 7th September 1919, she boarded the *Euripides* for Australia. During the war years she had been a faithful correspondent, not just with her family in Maitland but also with her husband, Edye Stobo. Some of the last entries in her attestation 'papers' concern queries from him as to her current location. An urgent telegram had followed a letter sent on 5th February to the Base Records Office, Department of Defence, Melbourne.

> "Sir, Would you kindly inform me of the present location of Sister Louisa Stobo – my wife.
> I have always heard from her regularly by mail or cable until recently and there is probably a delayed letter (or two) on the way, and she may also be on her way to Australia.
> Thanking you for previous intimations received, the last one being dated 30-4-18, as to award of the Royal Red Cross decoration to my wife, and thanking you, in anticipation, for reply."

The Officer in charge at Base Records duly replied, saying he had no recent knowledge of Louisa, but it could be assumed that she was still with her unit. He promised to provide early advice when she was on her way back. It was not until late in the year, however, that Edye was to receive a cable, informing him she was on her way back to Australia.[379]

The *Mercury* reported Louisa's return, with Edye, to Maitland on 25th October 1919. A large number of relatives and friends were at the station, along with representatives of most of the local patriotic bodies who had supported her efforts throughout the war to provide small 'comforts' to the soldiers she encountered. Louisa thanked the group for their kind welcome but especially extended thanks to the people of Maitland for their support of sick and wounded soldiers. She had distributed the money received to the best of her ability, she told them and there were cheers as she left for a family reunion at Oakhampton.

A more formal occasion, reported in full in the *Mercury*, was held several

379. 'Sister Stobo RRC,' Maitland Daily Mercury, 2 October 1919, p. 4
380. 'Sister Stobo's Return,' Maitland Daily Mercury, 27 October 1919, p. 4

days later, to honour both Louisa and Sisters Godfrey[381] and Lackey[382] who had also returned from the war.[383] The gathering was held by the Oakhampton Nurses' Comforts Fund committee and while three Sisters were honoured, only two could be present. The third, Sister Lackey, was unable to attend.

Among the large number of guests, there were returned soldiers - Lieutenant Mears and Privates Bergin, Hotchkiss, Miller, Finney and Creighton, who praised the Sisters and hailed 'The Rose of No Man's Land' as they referred to them. Major-Chaplain A. S. McCook spoke of the dangers the women had faced and the work of the Sister Stobo Fund. Various dignitaries spoke, including a representative of the War Chest Committee, who detailed the origins of the Fund and letters of appreciation they had received from Sister Stobo. The guests cheered and sang, 'For they are Jolly Good Sisters', and commemorative brooches and bouquets were presented to the two ladies. Louisa thanked the committee for the work they had done in forwarding "parcels of wearing apparel and many dainties, which relieved the monotony of the camp fare."

She spoke of the generosity of the Maitland Patriotic Societies, in forwarding money for the relief of Maitland men, and gave many instances of how it had been distributed. She also related how Sister Godfrey was the subject of a painting by "a well-known English artist" to represent the 'Digger's Nurse,' and of how this had appeared in many English journals. Sister Godfrey, in her turn, thanked the committee for the way the comfort fund had looked after her. Her warm garments had been the envy of less fortunate colleagues, she said. Her mother spoke, praising the committee for looking after her daughter through the Oakhampton Nurses Fund, which she believed was the only one, outside of Sydney.

These celebrations at Maitland must have been a joy and a relief. Louisa's next few years would see her appointment as Matron at Maitland Hospital, a posting which was not to end as happily. The intense pressures of war, where she received one of the highest decorations conferred on nurses, were replaced by peacetime work among people who, perhaps, had not experienced that pressure and did not appreciate those imperatives. The next

381. Leila Bowie Godfrey, was born c.1892, in East Maitland
382. Ethel Janet Lackey, was born 1885, in East Maitland
383. 'Nurses Honoured, Returned Army Sister,' Maitland Daily Mercury, 3 November 1919, p. 6

decade also saw the death of her husband and the renewal of an older relationship, with more nuptials.

And what of her diary, abandoned so abruptly on 15th June 1915, only a few weeks before Robert Scobie's death.

In September 1919, there is a final page of entries, in the same flowing but abbreviated script. The four entries record Louisa's departure for Australia on the S.S. *Euripides* and appear to be a record of which Sisters were on board and other brief notes about patients. Some names were repeated in the entry. The steamer left Plymouth with 2000 troops on board.[384] Among them was Brigadier-General T.A. Blamey.[385] It arrived at Woolloomooloo docks a little over seven weeks later, on 24th October.[386] It was almost five years since Louisa had left.

Louisa received the Royal Red Cross First Class for her services during WW1.
This example Collection of Grant Thompson.

384. General Cable News, Newcastle Morning Herald and Miners' Advocate, 9 September 1919, p.5
385. 'Personal,' Daily Advertiser, Wagga Wagga, 8 September 1919, p. 2
386. 'Euripides Troops,' Daily Telegraph, 24 October 1919, p. 5

September 1919

Entries for 16 July 1915 and 6-9 September 1919

6.9.1919

Left Paddington London for Plymouth for S.S. *Euripides* ... beautiful day.

Sisters Wilson, Lackey, Campbell, Kirkham,[387] Green,[388] Palmer, Ferguson, Coleman, Campbell, Pierce, B. Hooper, Lawrence, M E (?), S/N Henderson, [389] Patton, S/N Mead, McMurtrie, S/N Elizabeth, S/N McMurtrie,[390] Invalid malaria, S/N Kirkham, Not on duty

7.9.1919 Sunday

Sailed 9 a.m. from Plym. (Plymouth)

8.9.1919

Calm bright days, windy p.m. 14 patients in Hosp. (hospital) 8 p.m. V.D. downstairs.

---

387. Laurie Kirkham was born c. 1890 in Ashbourne South Australia.
388. Possibly Doris Marion Green who was born in 1889 in Colac Victoria
389. Isabel Henderson was born c. 1892 in Tasmania
390. Elizabeth Scouler McMurtrie was born in Victoria

9.9.19

Calm sunny morning. Sisters Hooper and Green on duty.

....

These brief bright notes mark the beginning of Louisa's journey home

Index

R

S